An Introduction to Number Theory
Part I

Professor Edward B. Burger

THE TEACHING COMPANY ®

PUBLISHED BY:

THE TEACHING COMPANY
4151 Lafayette Center Drive, Suite 100
Chantilly, Virginia 20151-1232
1-800-TEACH-12
Fax—703-378-3819
www.teach12.com

ISBN 1-59803-421-9

Edward B. Burger, Ph.D.

Professor of Mathematics
Department of Mathematics and Statistics, Williams College

Edward B. Burger is Professor of Mathematics in the Department of Mathematics and Statistics at Williams College. He graduated summa cum laude from Connecticut College in 1985, earning a B.A. with distinction in Mathematics. In 1990 he received his Ph.D. in Mathematics from The University of Texas at Austin and joined the faculty at Williams College. For the academic year 1990–1991, he was a postdoctoral fellow at the University of Waterloo in Canada. During three of his sabbaticals, he was the Stanislaw M. Ulam Visiting Professor of Mathematics at the University of Colorado at Boulder.

Professor Burger's teaching and scholarly works have been recognized with numerous prizes and awards. In 1987 he received the Le Fevere Teaching Award at The University of Texas at Austin. Professor Burger received the 2000 Northeastern Section of the Mathematical Association of America Award for Distinguished College or University Teaching of Mathematics and the 2001 Mathematical Association of America's Deborah and Franklin Tepper Haimo National Award for Distinguished College or University Teaching of Mathematics. In 2003 he received the Residence Life Academic Teaching Award at the University of Colorado. Professor Burger was named the 2001–2003 George Polya Lecturer by the Mathematical Association of America and in 2004 was awarded the Chauvenet Prize—the oldest and most prestigious prize awarded by the Mathematical Association of America. In 2006 the Mathematical Association of America presented him with the Lester R. Ford Prize, and he was listed in the *Reader's Digest* annual "100 Best of America" special issue as "Best Math Teacher." In 2007 Williams College awarded Professor Burger the Nelson Bushnell Prize for Scholarship and Teaching; that same year, he received the Distinguished Achievement Award for Educational Video Technology from The Association of Educational Publishers. Professor Burger's research interests are in number theory, and he is the author of 12 books and more than 30 papers published in scholarly journals. He coauthored with Michael Starbird *The Heart of Mathematics: An invitation to effective thinking*, which won a 2001 Robert W. Hamilton Book Award. They also coauthored a general audience trade book titled *Coincidences, Chaos, and All That Math Jazz*.

This is Professor Burger's third course for The Teaching Company. He previously taught *Zero to Infinity: A History of Numbers*, and he also co-taught *The Joy of Thinking: The Beauty and Power of Classical Mathematical Ideas*.

In addition, he has written seven virtual video textbooks on CD-ROM with Thinkwell and has starred in a series of nearly 2,000 videos that accompany the middle school and high school mathematics programs published by Holt, Rinehart and Winston.

Professor Burger has served as chair of various national program committees for the Mathematical Association of America; he serves as associate editor of the *American Mathematical Monthly,* and he is a member of the board of trustees of The Educational Advancement Foundation.

Professor Burger is a renowned speaker and has given more than 400 lectures around the world. His lectures include keynote addresses at international mathematical conferences in Canada, France, Hungary, Japan, and the United States; mathematical colloquia and seminars at colleges and universities; presentations at primary and secondary schools; entertaining performances for general audiences; and television and radio appearances including WABC-TV, the Discovery Channel, and National Public Radio.

Acknowledgments

Lucinda Robb had been encouraging me to return to The Teaching Company classroom for nearly two years. I want to sincerely thank her for her cheerful patience and constant enthusiasm. If it were not for her encouragement and support, I would not have had the wonderful opportunity to create this course and its historical counterpart, *Zero to Infinity: A History of Numbers*.

I also wish to express my sincere appreciation to the *Introduction to Number Theory* team at The Teaching Company, who made the entire process—from preproduction through postproduction—so pleasurable. Marcy McDonald provided excellent editorial suggestions and comments about the course structure. Zach "Zax" Rhoades was an outstanding producer who beautifully integrated the lectures with the visual elements. Tom Dooley and Jim Allen were the technical masterminds performing magic in the control room.

Within the world of mathematics, I wish to express my deepest gratitude to Professor Deborah J. Bergstrand from Swarthmore College. Professor Bergstrand provided invaluable and insightful suggestions that enhanced these lectures. Her contributions and dedication to this project were spectacular, and I thank her for all her efforts and for her friendship. From The University of Texas at Austin, I wish to thank my Ph.D. advisor, Professor Jeffrey D. Vaaler, who was the first to show me the beauty, wonder, and mystery hidden within the world of number theory.

Table of Contents
An Introduction to Number Theory
Part I

An Introduction to Number Theory

Scope:

The study of numbers—an area of mathematics now known as *number theory*—dates back to antiquity. Through the intervening millennia, creative and curious people around the world have pondered the meaning and nuance of numbers. Today, number theory is one of the exciting and active branches of modern mathematics that have seen great breakthroughs, such as Fermat's Last Theorem; great applications, such as public key cryptography; and great open questions, such as the Riemann Hypothesis—a complete and correct proof of which would entitle its author to a million-dollar prize.

In this course we will begin at the beginning and delve into the basic structure of numbers. We will then move into the study of surprising and stimulating results that have both tickled and confounded humankind for thousands of years. Beyond the ticker tape of numbers that possess clear and elegant patterns, we will explore the enigmatic prime numbers, discover the synergy between rational and irrational numbers, visit the world of algebraic and transcendental numbers, and journey into several modern areas including elliptic curves—a critical key to unlocking the proof of Fermat's Last Theorem after 350 years. While some mathematical confidence on the part of the student would be useful, these lectures will always paint a big picture of the main ideas in an accessible and nontechnical fashion before highlighting the intriguing and delicate details.

Lecture One
Number Theory and Mathematical Research

Scope:

In this opening lecture we will take our first steps into the abstract world of number theory and see how it fits within the larger mathematical landscape. We will come to view mathematics as a living and growing intellectual pursuit, and mathematicians both as artists creating new worlds and as explorers attempting to better understand our universe. Throughout their scholarly journey, mathematicians navigate through the subtle and narrow confines of truth. Mathematicians create *conjectures*—highbrow verbiage for educated guesses. Those conjectures are either proven to be true, in which case they are designated as *theorems*, or a counterexample is discovered demonstrating that the conjecture, in its generality, is false. If true, then the mathematical mind asks if the result can be extended or generalized. If false, then the mathematician wonders if the statement can be salvaged. Here we will describe this intellectual exploration and how it allows the frontiers of mathematical knowledge to move forward. Number theory is one of the oldest and most important branches of mathematics. At its very essence, number theory is the study of the natural numbers: 1, 2, 3, 4, and so forth. More precisely, it is an intellectual discipline concerned with the arithmetical structure of the natural numbers and the extensions and generalizations of such numbers. We will introduce some of the main branches of number theory, foreshadowing the journey ahead. After this overview of the number theoretic ideas we will discover throughout our course, we will close with a proof of our first mathematical truth: the whimsical "theorem" asserting that every natural number is interesting.

Outline

I. Welcome to the world of number.

   A. The motivation for number theory.

   1. We begin with the counting numbers, which we call the *natural numbers*: 1, 2, 3, 4, 5, … .

   2. Numbers evolved from useful tools to objects of independent interest.

3. The true motivation behind the theory of numbers comes from a desire to study the simplest numbers: the natural numbers. One of the fundamental recurring themes throughout this course is that when we explore "simple" objects in great depth, we uncover otherwise invisible delicate structure.

B. Different types of numbers.
   1. The natural numbers: 1, 2, 3, 4, 5, … .
   2. The integers: … , −3, −2, −1, 0, 1, 2, 3, … .
   3. The ratios of integers (fractions) are called *rational numbers*.
   4. The numbers that are not ratios are called *irrational numbers* (numbers that are not rational).
   5. The decimal numbers are called *real numbers*, which can be viewed as points on a number line.

C. The surprising synergy between numbers.
   1. Once we identify different types of numbers, they take on a life of their own, dictated by the laws of nature and mathematics, and we can study their rich and intricate personalities.
   2. We will see many surprising connections between these different numbers. This interplay and synergy between numbers will be another recurring theme throughout our course.

II. The culture of mathematics.
   A. Mathematics is an ever-growing area of active research.
      1. Mathematics is an abstract universe of both nature and the mind.
      2. Mathematicians are at once explorers and artists.
      3. Mathematics is a search for structure.
   B. Moving the frontiers forward.
      1. Most mathematics is not yet understood.
      2. Mathematicians, as a community, build on each other's works to move the boundaries of our understanding outward.
      3. We will begin with some basic self-evident truths—known as *axioms*—and build upward.
      4. We will then study simple objects and search for patterns.

5. Patterns lead to conjectures, which in turn can lead to theorems.

6. New theorems are discovered, proofs are created, and then the mathematics community reviews and accepts the new results.

C. The power of proof.

    1. Mathematics is built on the notion of rigorous proof.

    2. Mathematicians are at once artists and lawyers.

    3. Our course will celebrate the creative art of rigorous proof.

III. What is number theory?

    A. Analytic number theory: a focus on primes.

        1. The prime numbers are those natural numbers greater than 1 that cannot be written as a product of two smaller natural numbers.

        2. The first few prime numbers are 2, 3, 5, 7, 11, 13, and 17.

        3. The prime numbers form the building blocks for all natural numbers: Every natural number greater than 1 is a product of prime numbers.

        4. Are there infinitely many prime numbers?

        5. The prime number theorem, proved in 1896, in some sense reveals what proportion of natural numbers are primes: As we consider larger and larger values of n, the number of primes up to n approaches the quantity $n/(\log n)$. One of the most important open questions in mathematics asks if this result can be improved. The question is now known as the *Riemann Hypothesis*.

        6. This area of number theory is known as *analytic number theory* because it employs the techniques of calculus to establish the truth of its results.

    B. Algebraic number theory: a focus on arithmetic.

        1. Suppose we have an equation that just involves addition, subtraction, and multiplication of natural numbers. Can we always find natural numbers that are solutions? Such equations are called *Diophantine equations*.

        2. For example, if we consider the equation that arises from the Pythagorean Theorem: $x^2 + y^2 = z^2$, then we see that $x = 3$, $y = 4$, and $z = 5$ are natural numbers that give a solution. Are there others? Infinitely many?

3. As we will see, the famous "Fermat's Last Theorem" states that a related equation has no natural-number solutions. Finding a complete proof of this assertion remained one of the most prized open questions in mathematics for over 350 years until it was finally established in the mid-1990s.

4. The study of solutions to these types of equations will lead us to discover a generalized notion of integers and then a generalization of the prime numbers.

5. Because this area of number theory is inspired by a search for solutions to equations, it is known as *algebraic number theory*.

6. Given that both analytic and algebraic number theory involve the study of primes and their generalizations, it will not be surprising to see a synergistic interplay between these two branches of the theory of numbers.

IV. The vistas ahead in this lecture series.

   A. Elementary number theory.

      1. We start by discovering some interesting patterns involving natural numbers. These attractive patterns of numbers not only hold an independent aesthetic appeal, but they will also be extremely useful tools as we move to deeper areas of number theory.

      2. These early explorations into pattern will also provide us with the opportunity to craft our own conjectures and to become accustomed to the world of rigorous proof.

   B. Analytic number theory will formally introduce the prime numbers and study their central role in the universe of number.

   C. Modular arithmetic.

      1. Combining the properties of primes with the "clock arithmetic" of cycles, we will discover a world of arithmetic involving division that focuses on remainders rather than quotients.

      2. As we will discover, this classical study holds one of the most important modern applications of number theory: public key cryptography.

   D. Algebraic number theory. In our attempt to find solutions to certain equations, we will come upon parallel universes of number that will lead us to rethink the basic mathematics we were taught in school.

E. Algebraic geometry.
    1. Here we will combine the power of algebra and geometry to discover an important connection between solutions to certain equations and points on certain curves.
    2. The interplay between number theory and geometry is one of the most profound elements of modern number theory.

F. Algebraic and transcendental numbers.
    1. Here we will explore whether there are any other numbers beyond those that are the solutions to the equations we studied in algebraic number theory.
    2. This question remained unanswered for thousands of years until the definitive answer was finally found in the mid-1800s, which in the scope of mathematical history was "yesterday."

G. Continued fractions.
    1. Beyond writing numbers as familiar decimals, we will discover an alternative manner to express numbers that is much more sympathetic to number theory inquiries.
    2. This manner of expanding numbers allows us to build insights into many phenomena, from why 22/7 is so close to π, all the way to why we have a 12-note chromatic musical scale.

H. Throughout the course, we will offer applications and stories both famous and whimsical that will not only enhance the number theory at hand but will provide a context within which we can better appreciate our discoveries about the structure within the world of number.

V. Every number is interesting.

  A. A personal passion for number theory.
    1. The intrigue of number—a notion that is at once basic and profound.
    2. The creativity and originality involved in crafting a proof.
    3. Painting an abstract portrait of beauty and detail.
    4. Viewing mathematical proof as a new form of art.

B. "Theorem": Every natural number is interesting.

    1. The "proof" of this whimsical assertion illustrates an argument known as *mathematical induction*.

    2. We consider the first natural number, 1. It is the smallest natural number, and thus it is certainly interesting.

    3. The next natural number, 2, is the smallest even number, which is interesting.

    4. Could there be a natural number that is *not* interesting? If so, then there must be a smallest one; that is, the smallest natural number that is not interesting. In other words, every natural number less than this number *is* interesting, and this number is the first natural number that is *not* interesting—but isn't *that* interesting? This exclamation concludes our humorous proof.

    5. The world of number theory is indeed interesting. Our journey ahead is teeming with deep ideas, profound insights, and incredible discoveries. While we will see serious mathematics revealed in details and proofs, we will always place those technical points in context within the panorama of number.

Questions to Consider:

1. Give several examples of situations in which logical proofs are required.

2. Returning to the whimsical mathematical proof that every natural number is interesting, suppose that all the numbers from 1 to 20 are previously known to be interesting. How can you use this fact to show that the number 21 *must* also be interesting?

Lecture One—Transcript
Number Theory and Mathematical Research

The study of numbers, an area of mathematics now known as *number theory*, dates back to antiquity. Through the intervening millennia, creative and curious people around the world have pondered the meaning and nuance of numbers.

Today, number theory is one of the exciting and active branches of modern mathematics that have seen great breakthroughs such as Fermat's Last Theorem; great applications such as public key cryptography, and great open questions such as the Riemann Hypothesis—a complete and correct proof of which would entitle its author to a million-dollar prize.

In this course, we'll begin at the very beginning and delve into the basic structure of numbers, and then move into the study of surprising and stimulating results that have both tickled and confounded humankind for thousands of years.

Beyond the ticker tape of numbers that possess clear and elegant patterns, we'll explore the enigmatic prime numbers. We'll discover the synergy between rational and irrational numbers. We'll visit the world of algebraic and transcendental numbers, and we'll journey into several modern areas, including elliptic curves—a critical key to unlocking the proof of Fermat's Last Theorem, over 350 years of work.

While some mathematical confidence would be helpful for our lectures, we'll always paint a big picture of the main ideas in an accessible and nontechnical fashion before celebrating the intriguing and delicate details.

In this opening lecture, we'll take our first steps into the abstract world of number theory and see how it fits in within the larger mathematical landscape. We'll come to view mathematics as a living and growing intellectual pursuit, and mathematicians as both artists—creating new worlds—and as explorers—attempting to build better understanding of our universe.

Throughout their scholarly journey, mathematicians navigate through the subtle and narrow confines of truth. Mathematicians create *conjectures*, which is really just highbrow verbiage for educated guesses.

These conjectures are either proven to be true, in which case they're awarded the title *theorem*, or a counterexample is discovered, demonstrating that the conjecture, in its generality, is actually false. If true, then the mathematical mind asks if the result can be extended or generalized. If false, the mathematician wonders if the statement can be salvaged—that's how the frontiers move forward.

Here we'll describe this intellectual exploration and how it allows the frontiers of mathematical knowledge to genuinely move forward. Number theory is one of the oldest and most important branches of mathematics.

At its very essence, number theory is the study of the *natural numbers*—the numbers 1, 2, 3, 4, and so on. More precisely, it is an intellectual discipline concerned with the arithmetical structure of the natural numbers and the extensions and generalizations of these numbers.

Now, while many people may believe that all of mathematics is understood, in this course we'll see that mathematics in general, and number theory in particular, are ever-evolving works in progress and, just as with other artistic pursuits, are driven by the fashion and tastes of today's practitioners.

In this lecture, we'll offer a window into the mathematical culture and then introduce some of the main branches of number theory, foreshadowing our journey ahead.

After this overview of number theoretic ideas that we'll encounter throughout our course, I thought we'd close with a proof of our first mathematical truth—the whimsical "theorem," if you will, asserting that every natural number is, in fact, interesting.

Well, as we enter the world of numbers, we first consider the motivation for number theory. We begin with the counting numbers, which we call the natural numbers. They're the numbers 1, 2, 3, 4, and so forth, and they will follow us throughout our entire journey.

Now, numbers slowly evolved from useful tools to objects of independent interest. If you think about it, numbers at the beginning were used to describe things; they were adjectives. We have 5 apples. We have 10 sheep. But soon humankind and the human imagination was captured, and one isolated this adjective and turned it into a noun—the number 3, or the number 10. And once it became a noun, it became an object of curiosity and an object of study.

The true motivation behind the theory of numbers comes from a desire to study the simplest numbers—these natural numbers. One of the fundamental recurring themes throughout our entire course is that when we explore "simple" objects with great depth, we uncover otherwise-invisible delicate structure, and we'll see this again and again.

There are many different types of numbers that we'll see throughout our course, and I wanted to quickly introduce you to them, although we'll see them in greater depth as we go. The natural numbers, of course, are the numbers 1, 2, 3, 4, and so on. The *integers* are the natural numbers together with their negatives and with 0, so 1, 2, 3, 4, and so on; together with −1, −2, −3, and so on; and we also toss in 0. Those are the integers.

The ratios of integers, which we can think of as fractions, are called the *rational numbers*. The numbers that are not ratios are called *irrational numbers*—numbers without ratios, so to speak, and these confounded mathematicians way back into the ancient Greek time of Pythagoras, as we'll see.

The decimal numbers are called the *real numbers* and can be viewed as points on a number line from our youth. Or the numbers that pop up when we use a calculator and see the decimal point and lots of digits.

Well, once we identify different types of numbers, they do genuinely take on a life of their own, dictated by the laws of nature and mathematics. Much of the journey ahead, in fact, is a thorough study of their rich and intricate personalities, and I'll actually refer to numbers as having personalities, although we'll mean that in a mathematical context. In fact, we'll see many surprising connections between these different types of numbers. This interplay and synergy between numbers will be another recurring theme throughout our course.

Before moving on, I wanted us to take a moment to reflect upon the broader culture of mathematics. Mathematics is an ever-growing area of active research. Mathematics is an abstract universe of both nature and the mind. For example, we see objects moving in our everyday lives, we see the motion of planets and now understand motion through the abstract ideas of calculus—the study of change. We see shapes and we wish to describe them, and this leads us to the abstract universe of geometry. We see quantities of objects and we count, and this leads us to the abstract world of number theory.

Mathematicians are at once explorers and artists. We see the world, and we try to capture—not on canvas—what we see, but we try to capture it through mathematical formulas and theorems.

Mathematics is a search for structure. When we see a pattern in our world, that is often a sign that in fact there's some underlying structure at work. So our quest in mathematics is to identify this underlying structure that explains the patterns we see.

Once we develop the structures and explain the patterns in the real world, we often find new abstract structures and patterns within the mathematics itself, which may have obvious connections with our original inspiration.

I now want to say a word about how we move the frontiers of mathematics forward. First and foremost, mathematics, and most of mathematics, in fact, has not yet been discovered, and even more is not totally understood. Mathematics is difficult, first of all, and creating new mathematics is, in fact, particularly challenging.

Mathematicians, as a community, build on each other's works to move the boundaries of understanding outward. The discipline of mathematics rests upon the foundations of a handful of basic self-evident truths known as *axioms* and then rises up from there.

We study simple objects or phenomena and search for patterns. Now, the patterns that we discover lead to our conjectures, which in turn, in the best of all possible worlds, lead to new theorems. Well, once a new theorem is discovered and a rigorous mathematical proof is created, the work is submitted to a scholarly journal, and the mathematics community reviews the work, usually by sending it out to anonymous referees who are also experts in the area.

If the work is correct and complete, and of sufficient interest, then it is accepted into the journal, and it's accepted by the community, and the new results are deemed a theorem.

Thus, in the final analysis, the pillars of mathematics are the rigorous proofs. So what's a rigorous proof? Well, a rigorous proof really is just a string of declarative assertions, each deduced from the one before. So in some sense, mathematicians are at once artists and lawyers—we're trying to create an argument. But unlike the lawyer for whom guilt and innocence rests on a reasonable doubt, here there is no reasonable doubt. It has to be ironclad.

Our course will continually celebrate the creativity involved in constructing a rigorous proof, and we'll actually see rigorous proofs for ourselves and construct them and create them, and we'll appreciate the creativity involved that much more.

Well, stepping back now from the panoramic view of all of mathematics, we now focus on the question, what is number theory? Well, the two main branches from which most of the questions of number theory fall are known as *analytic* and *algebraic* number theory.

Analytic number theory focuses on the primes. Now, the prime numbers, which we'll say much more about in future lectures, are those natural numbers greater than 1 that cannot be written as the product of two smaller natural numbers. So, for example, the first few prime numbers are 2, 3, 5, 7, 11, 13, 17. They can't be broken down into smaller parts by multiplication. Six, however, is not a prime, because 6 can be written as 2×3—two smaller natural numbers.

We'll study Euclid's insight that the prime numbers form the building blocks for all natural numbers; in particular, every natural number greater than 1 is a product of prime numbers. Are there infinitely many prime numbers? Euclid answered this question, and his proof is considered to be one of the most elegant arguments in all of mathematics, and we'll see this for ourselves.

Refining the issue of the infinitude of primes, mathematics wondered about the proportion of natural numbers that are prime. As we'll see, this inquiry led to the discovery of what's called the *prime number theorem*, proved in 1896. The prime number theorem gives an approximation to how many primes we have up to any particular point.

One of the most important open questions in mathematics asks how good this approximation truly is. In its precise form, this question is now known as the *Riemann Hypothesis* and remains one of the most important open questions in all of mathematics.

This area of number theory is known as *analytic number theory*, because it often employs the techniques of calculus to establish the truth of its results. Now remember, calculus, let me just remind you, is the study of growth—the study of change, the study of movement. So studying the growth of primes using the mathematical study of growth seems appropriate, and the techniques, in fact, of calculus, are used. Therefore it's called *analytic number theory* because the area of mathematics with which the calculus resides is known as *analysis*.

Algebraic number theory, on the other hand, has its focus on arithmetic. Here we'll take a look at equations that just involve addition, subtraction, and multiplication of natural numbers. The most basic question we can ask is, can we always find natural numbers that are solutions—namely, natural numbers that will satisfy the equation, or make the equation true?

Such equations are called *Diophantine equations*, named after the great Greek mathematician Diophantus of Alexandria. For example, if we consider the equation that arises from the Pythagorean Theorem, that equation that we saw quite a while ago in our grade school days, $x^2 + y^2 = z^2$, remember that x^2 just means $x \times x$. Then we see that if x takes on the value 3, if $y = 4$, and $z = 5$, well, these are natural numbers, and notice, they satisfy this equation. For if we take 3 and we square it, that's 3×3; that gives us 9. If we take 4 and we square it, that's 4×4; that's 16. And $9 + 16$ is 25, which notice is indeed 5^2. So we have a solution, and in fact a solution where all the unknowns are natural numbers. Well, are there other solutions? Are there infinitely many? These are the questions that we face.

As we'll see, the famous "Fermat's Last Theorem" states that a generalized version of the Pythagorean equation has no natural-number solutions. Finding a complete proof of this assertion remained one of the most prized open questions in mathematics for over 350 years, until it was finally established in the mid-1990s. This actually indicates and provides us with a wonderful life lesson that number theory and mathematics teaches us, which is that we should really always be very, very tenacious and never give up. Tenacity is quite often the best way to combat difficult, challenging issues. Never, never give up, because in fact the solution might be right around the corner, and we'll see this theme occurring in our course.

The study of solutions to these types of equations will actually lead us to discover a generalized notion of integers, and in turn, a generalization of the prime numbers. Well, because this area of number theory is inspired by a search for solutions to equations, it is known as *algebraic number theory*, because we use the algebra that we're used to seeing from our youth to look at solutions.

Given that both analytic and algebraic number theory involve the study of primes and their generalizations, it will not be surprising to discover an important interplay between these two branches of the theory of numbers. We'll see that interplay for ourselves.

Now I want to offer a few vignette scenes of the vistas ahead in our journey through number. We'll open with an excursion into elementary number theory. *Elementary*, here, in mathematics, doesn't mean "easy." It means that the questions can be posed and could be understood by many people, and the solutions don't involve very advanced mathematics but instead a very clever coupling of ideas together.

We'll discover some interesting patterns involving the natural numbers. These attractive patterns not only hold an independent aesthetic appeal but are also extremely important tools required to understand the deeper ideas of number theory to come. So we will be building the foundations.

In addition, these early expeditions into pattern will provide us with the opportunity to create our own conjectures and to gently introduce us to the world of rigorous proof. In some sense, we'll enter the world of rigorous proof at the shallow end and slowly move into the deeper, more interesting end of the number theory pool.

We'll then turn to analytic number theory, and here we'll formally introduce the prime numbers and study their central role in the universe of number.

Next, we'll see a different method of calculation known as *modular arithmetic*. Modular arithmetic is really nothing more than "clock arithmetic." For example, if we add 3 hours to eleven o'clock, our clock would show two o'clock. So we somehow came around after twelve and started over again. This is the arithmetic of cycles.

Combining the properties of primes with the clock arithmetic that we'll develop, we'll discover a world of arithmetic involving division that focuses on remainders rather than quotients.

As we'll see, this classical study holds one of the most important modern applications of number theory—public key cryptography, which is the fundamental and foundational feature that occurs in all modern digital communication. We use it all the time.

This will naturally bring us to algebraic number theory, whereby searching for simple solutions to certain equations will come upon parallel universes of number that will cause us to rethink the basic arithmetic and mathematics that we were taught in school.

Next, we'll see the potent combination of algebra and geometry as we explore the modern area known as *algebraic geometry*. Here we'll discover an important connection between solutions to certain equations and points on certain curves. The interplay between number theory and geometry is one of the most profound areas of current research in number theory today.

We'll then move to the mysterious world of algebraic and transcendental numbers. Here we'll wonder if there are numbers beyond those that are the solutions to the equations we considered in algebraic number theory, and we'll see the answer is yes. This issue remained unresolved for thousands of years until the definitive answer was finally found in the mid-1800s, which in the scope of mathematical history was essentially "yesterday."

We'll then close the course by considering how we can represent numbers. Beyond writing numbers as familiar decimals, we'll discover an alternative manner to express numbers that is much more sympathetic to number theory inquiries.

This manner of expanding numbers allows us to build insights into many phenomena: from why 22/7 so closely approximates the number π, and then extends to why we have a 12-note chromatic musical scale. So we'll see many, many features that come clear by just looking at how we represent things.

Throughout the course, we'll offer applications and stories, both famous and whimsical, that will not only enhance the number theory at hand but will provide a context within which we can better appreciate our discoveries within the world of number.

Before closing this introduction, I wanted to make a few personal remarks. I'm a number theorist, and therefore I'm particularly excited about the prospect of sharing this incredible universe of number with you. In my research, I'm actually proving new theorems in number theory, and so this is something that's near and dear to my heart.

Number theory, in fact, really is one of my intellectual passions. For one thing, the questions are often very easy to pose, and yet the questions involve answers that might elude humanity for thousands of years.

I believe that numbers are intriguing objects of nature and our imagination. They're at once basic and profound. I'm also drawn to the creativity and originality involved in crafting arguments—that is, rigorous proofs that we'll see again and again.

The proofs that we'll see throughout our course will together paint an abstract portrait of beauty and detail. Now, there will be moments of technical detail, some of which will have algebraic components. There's a wonderful quote due to my late friend, the 20th-century mathematician Gian-Carlo Rota, who spent most of his career at MIT. Here's what he said about mathematics: "A mathematician's work is mostly a tangle of guesswork, analogy, wishful thinking, and frustration, and proof, from being the core of discovery, is more often than not a way of making sure that our minds are not playing tricks on us." So, a wonderful quote that really captures the spirit of what we do in mathematics.

Number theory can be frustrating, but I urge you not to get bogged down in the details if that's not of interest to you. Instead, I urge you to take it all in at once and let the ideas leave you with an impressionistic sense of the arguments. In fact, I propose that we should view mathematical proofs as a style of art, perhaps a style with which many are not familiar.

When we first look upon the works, they look jumbled, confusing, abstract, and challenge our senses. I urge you to stay the course. Just as with any new type of art, we can only appreciate the style once we've had a chance to see many examples and know where to look and what to see. So please think of me as your number theory museum curator.

I thought that we should really leave the lecture, this first, opening lecture, by actually proving something. I thought that we should prove a theorem that captures the spirit of what we are going to be looking at for the rest of the lectures, even though it is a little bit whimsical.

The "theorem" is every natural number is interesting. How would you prove something like this? Well the "proof" of this whimsical assertion actually illustrates a style of argument known as *proof by contradiction*. Here we actually will assume the opposite of what we hope to prove and hope to deduce something that's just totally false, thereby contradicting ourselves and thus showing that our assumption was genuinely false.

So now let's take a look at why every natural number is interesting. We consider the first natural number, which is the number, of course, 1. It's the smallest natural number, and thus that's certainly interesting. It's the first one, that's certainly interesting. So 1 is interesting.

The next natural number is 2, and notice, that's the smallest even natural number, which is certainly interesting. Also, by the way, 2 is the first prime number, which is really interesting. So we see that both 1 and 2 are indeed interesting.

Now could there be a natural number that is *not* interesting? Well, if so, then there must be a smallest such number—that is, the smallest natural number that is not interesting. In other words, every natural number less than this one *is* interesting, and this number is the first natural number that is *not* interesting. Well, isn't *that* interesting?

So, the point is that if you think you have the smallest natural number that's not interesting, then since it's the first one, that automatically means it's interesting. That shows us that we never have a number that's not interesting, because in fact whenever we have one, the first one would be interesting, and it skyrockets from there.

This exclamation concludes our humorous proof. But it really does illustrate something important. The world of number theory is indeed interesting, and every number is worthy of exploration. Our journey ahead is teeming with deep ideas, profound insights, and incredible discoveries.

I'm delighted we will share the world of number theory together.

Lecture Two
Natural Numbers and Their Personalities

Scope:

The most natural place to begin our journey into number is with the numbers we have always counted on—the natural numbers: 1, 2, 3, 4, and so forth. In this lecture we celebrate the main characters of our story and foreshadow the methods by which we uncover and establish the truth of theorems. Of course, there are infinitely many natural numbers, and making that reality intuitive, even today, is much more challenging than it first appears. As we will discuss, most numbers, for example, cannot even be named. But can *all* numbers be understood in either practice or in theory? Here we will highlight the remarkable reality that mathematical thinking allows us to verify truths about *all* numbers—the unending collection of values, most of whom we will never name, imagine, or comprehend. This reality will underscore the power of rigorous proof. Once we introduce the main characters of this course, we can move to the means by which we will study their personalities; these traits arise through the introduction of arithmetic. It is this marriage of number and arithmetic that gives birth to notions of number theory. We will close this lecture with some basic but striking arithmetical observations—some, we will see, are provable theorems, while others remain hidden in a veil of mystery.

Outline

I. Speaking the language of numbers.

   A. Natural numbers are the numbers in the collection {1, 2, 3, 4, ... }.

   B. Integers consist of the natural numbers together with their negatives and zero.

   C. Rational numbers are all ratios of integers. Specifically, a rational number is a number that can be expressed as a/b, in which a is an integer and b is a natural number.

   D. Real numbers are all "decimal numbers," that is, the numbers that correspond to points on the number line.

II. Do large numbers have any intuitive meaning to us?

    A. The endless stream of natural numbers.

        1. Clearly there are infinitely many natural numbers, and our goal in number theory is to attain a deep understanding about these objects.

        2. The desire to understand an infinite collection of objects in a finite amount of time is the fundamental feature that makes this inquiry both subtle and abstract—and thus generates much of its intrigue.

    B. Naming the numbers.

        1. From a practical point of view, our understanding of numbers is dictated by their utility in our everyday lives.

        2. Some early cultures counted "1, 2, many." Today we are familiar with millions and trillions.

        3. As number theory enthusiasts, we wish to study all of the natural numbers, so naming them becomes an issue. Even silly names can be useful, such as "googol" and "googolplex," the origins of which are amusing stories.

        4. However, there are infinitely many numbers, and most of them—from the point of view of language—have not been named.

    C. How large is our universe?

        1. Not all quantities within our universe can be named.

        2. The size of our universe is approximately 10^{79} atoms.

        3. Skewes number, $10^{10^{10^{34}}}$, is believed to be the largest number that appears in a significant mathematical theorem.

        4. While these numbers are so large that they might have no intuitive meaning to us, such sizes are insignificant from a number theory prospective because there are infinitely many numbers greater than these relatively tiny values.

        5. Here in this field of inquiry we wish to study those objects that we, in practice, will never see—numbers whose names we'll never utter.

D. The power of proof.

    1. We wish to make discoveries about endless lists of numbers—we make conjectures and strive to show that they are indeed valid.

    2. We cannot simply check each natural number to verify a conjecture.

    3. We rely on logical and abstract thinking and craft a rigorous proof.

III. The personality of numbers.

  A. Searching for structure within the numbers.

    1. We explore the numbers and search for interesting patterns.

    2. Those patterns often lead to important insights into the structure of numbers.

    3. We look at patterns within the numbers 1, 4, 9, 16, ...

  B. The central notion of divisibility.

    1. We look at even and odd numbers and the patterns they exhibit.

    2. One important way of detecting the personality of a natural number is to factor it.

    3. The factors are also known as *divisors*.

    4. The notions of divisibility and divisors are the centerpieces of number theory.

  C. Products of consecutive even or odd numbers (plus 1).

    1. We examine a pattern within products of evens (plus 1).

    2. We prove the pattern and discover a theorem.

    3. We look at a similar pattern with odd numbers.

IV. Collatz's question.

  A. Introducing the famous "$3n + 1$ question."

    1. Given a natural number n, we use it to generate a list of numbers by the following process.

    2. If n is even, then the next natural number on our list is $n/2$; if n is odd, then the next natural number on our list is $3n + 1$.

    3. We now apply the same procedure with this new number to produce the next number on the list and continue.

    4. This sequence was first studied by Lothar Collatz in 1937.

B. Illustrations and examples.

    1. If we start with 1, we see: 1, 4, 2, 1, 4, 2, 1, … . Thus if we generate a 1, we know the rest of the list will just repeat: 1, 4, 2, 1, 4, 2, 1, and so forth.

    2. If we start with 2, we see: 2, 1, 4, 2, 1, and so forth.

    3. If we start with 3, we see: 3, 10, 5, 16, 8, 4, 2, 1, 4, 2, 1, and so forth.

    4. If we start with 4, we see: 4, 2, 1, 4, 2, 1, and so forth.

    5. If we start with 11, we see: 11, 34, 17, 52, 26, 13, 40, 20, 10, 5, 16, 8, 4, 2, 1, 4, 2, 1, and so forth.

C. Searching for a pattern: the known.

    1. It appears that no matter which we start with, the list always eventually becomes an endless run of 1, 4, 2, 1, 4, 2, 1, … .

    2. If we start with the modest number 27, the process will produce a list of 111 numbers before we finally see our first 1. Within those first 111 numbers we would see numbers as large as 9,232. But we do finally settle down to the familiar 1, 4, 2, 1, and so forth.

    3. Every natural number up to around 10^{18} has been used as the starting value. In each case, the list finally settles down to the repeating 1, 4, 2, 1, and so forth.

D. The endless unknown.

    1. Collatz conjectured that starting with *any* natural number, the list will eventually settle down to 1, 4, 2, 1, and so forth.

    2. This remains one of the most famous open questions in elementary number theory.

V. Discovering your power through number theory.

A. Powers of 2. Let's produce the first few powers of 2: $2^1 = 2$; $2^2 = 4$; $2^3 = 8$; $2^4 = 16$; $2^5 = 32$; $2^6 = 64$; … ; $2^9 = 512$; … ; $2^{24} = 16,777,216$.

B. A screed of digits.

    1. What are the first (left-most) digits we see in the above numbers? 2, 4, 8, 1, 3, 6, and then later, 5.

    2. Are there powers of 2 that begin with the missing two nonzero digits, 7 and 9?

3. One can check that $2^{46} = 70,368,744,177,664$ and $2^{53} = 9,007,199,254,740,992$.

4. Thus we see that every digit from 1 to 9 is the first (left-most) digit for some power of 2. Is there a power of 2 that begins with 10, or 11, or 12? Is there a power of 2 that begins with your social security number? How about your social security number, followed by your birth date, followed by your cell phone number (including the area code)?

C. A surprising theorem.

1. As amazing as it might at first seem, there is a theorem that asserts that given any natural number n, there exists a power of 2 whose left-most digits agree with the digits of the given natural number n.

2. The truly amazing aspect of this assertion is that this result can be proved for all natural numbers.

D. The ideas involved in the proof of the theorem.

1. At the very end of this course, we will have developed enough mathematical machinery to appreciate the ideas behind why this result is true.

2. As we will discover, to prove this result, we must travel beyond the world of natural numbers and employ results involving the irrational numbers. This connection will be our final illustration of the incredible synergy between the many different types of numbers.

Questions to Consider:

1. Consider the numbers 61 and 64, which are clearly 3 apart. In what ways do their *arithmetic* personalities differ?

2. Verify that the "$3n + 1$ question" has an affirmative answer if you start with 7.

Lecture Two—Transcript
Natural Numbers and Their Personalities

The most natural place to begin our journey into number is with the numbers we've always counted on—the natural numbers: 1, 2, 3, 4, and so on. In this lecture, we celebrate the main characters of our story and foreshadow the methods by which we'll uncover and establish the truth of theorems.

Of course, there are many, in fact infinitely many, natural numbers, and making that reality intuitive, even today, is much more challenging than it first appears.

As we'll soon realize, most numbers, for example, cannot even be named. But can *all* numbers be understood either in practice or in theory? Here, we'll highlight the remarkable reality that mathematical thinking allows us to verify truths about *all* numbers— the unending collection of values most of which we'll never name, imagine, or even comprehend. This reality will underscore the power of rigorous proof.

Once we introduce the main characters of this course, we can turn to the means by which we could study their personalities. These traits arise through the introduction of arithmetic. It is the marriage of number and arithmetic that gives birth to the notions of number theory.

We'll close this lecture with some basic but striking arithmetical observations. Some we'll see are provable theorems, while others remain hidden within a veil of mystery.

Let's begin by recalling the different types of numbers that will follow us through our course. We have the natural numbers, the numbers 1, 2, 3, 4, and so on. The integers consist of the natural numbers together with their negatives and also 0. The rational numbers are all ratios of integers: Specifically, a rational number is a number that can be expressed as a/b, in which a is an integer and b is some natural number. So for instance 2/3, 5/2, −7/8 are all examples of rational numbers.

The real numbers are the "decimal numbers" that we're used to seeing on our calculators and computers. In other words, the numbers that correspond to points on the real number line.

This course is a study involving the structure of numbers. So, before we even begin, we have to ask a fundamental question about the numbers themselves. Do large numbers have any intuitive meaning to us? Clearly, there are infinitely many natural numbers, and our goal in number theory is to attain a deep understanding of these objects. The desire to understand an infinite collection of objects in a finite amount of time is the fundamental feature that make this inquiry both subtle and abstract, and thus alluring.

From a practical point of view, our understanding of numbers is dictated by their utility in our everyday lives. In fact, some early cultures actually counted "1, 2, many," because that's all they needed. Of course today, we throw around words like "millions," "trillions" and barely even blink an eye.

As number theory enthusiasts, we wish to study all the natural numbers, so naming them one by one becomes an issue. Even silly names can be useful, such as googol and googolplex, the origins of which are actually amusing stories.

A googol is defined to be 10 raised to the 100th power. What does that mean? It's the number 1 followed by 100 zeros. It was the 20th-century American mathematician Edward Kasner who introduced this word around 1938. As the legend goes, Kasner asked his nephew, whose name was Milton, who was about eight years old at the time, what name he would give to a really huge number. Young Milton answered, "Googol."

Later Kasner defined a "googolplex" to equal 10 raised to the googol power—that is, 1 followed by a googol number of 0s. These words have captured the public's imagination, and they're often used in referring to extremely large quantities.

Even though a googolplex is dramatically large, the reality remains that there are infinitely many numbers, and thus most of them, from the point of view of language, have not been named. They are unnamed.

Even in our everyday world, around us, we see extremely large quantities. Scientists, for example, estimate that the number of atoms in our universe is approximately 10 raised to the 79th power. That's a 1 followed by 79 zeros—a very large number indeed.

In mathematics, there's a ridiculously large number, known as *Skewes number*, that had the folklore distinction of being the largest number that appeared in a serious theorem. Skewes number equals, and get ready for this, 10 raised to the power 10 to the 10 to the 34th. That's a 1 with a lot of 0s after it. And we'll actually describe the theorem that contains it in Lecture Nine.

While these numbers are so large that they have really no intuitive meaning to us, such sizes are really insignificant from a number theory perspective, because there are infinitely many numbers greater than these relatively tiny values. In the scheme of things, when we get to even a googolplex, we haven't made much progress in our list of numbers because infinitely many natural numbers still remain to be accounted for. So we've basically made no progress, even with these enormous numbers.

Here in this field of inquiry, we study those objects, most of which, in practice, we'll never see—numbers whose names we'll never utter, and yet we'll prove mathematical truths about these unnamed numbers. In this course, we'll be making discoveries about endless lists of numbers, we'll be making conjectures, and we'll strive to show that those assertions are indeed valid for all numbers. We can't simply just check each natural number to verify a conjecture, since that would require us to perform literally infinitely many tasks. Instead, we rely on logical and abstract thinking, and craft a rigorous proof that establishes the assertions for all numbers in general.

So where do these conjectures come from? As we suggested in the first lecture, number theorists explore numbers and search for interesting patterns, and we'll do this throughout our course. We'll constantly be looking for interesting patterns. Those patterns often lead to important insights into the structure of numbers. In fact, the great 20th-century British analytic number theorist G. H. Hardy once wrote, "A mathematician, like a painter or a poet, is a maker of patterns. If his patterns are more permanent than theirs, it is because they are made with ideas."

You can see here that Hardy seems to be putting a hierarchical structure, putting number theory and mathematics perhaps above the painters and the poets. I don't know if I necessarily agree with that, but we do share this common creative thread; namely, we are searching for and trying to create patterns.

For example, let's just consider the following list of numbers that I'm going to give you, and let's see what we see. So, here's the list of numbers that I'm thinking about: 1, 4, 9, 16. Let's just stop there. If we look at that list of numbers, we see that they're all natural numbers, and they're getting larger. Those are patterns that we observe. But what else can we say?

Upon closer inspection, we see that each number is, in fact, a perfect square. That means that each number is actually some natural number multiplied by itself. Notice that 1 is 1 squared, 4 is 2 squared, 9 is 3 squared, 16 is 4 squared. So there's a pattern hidden in this list of numbers, and notice that this pattern involves products of smaller numbers—namely, products of numbers with themselves.

This example involving multiplication brings us to the most central and important notion in number theory, that of divisibility. For example, if a number is evenly divisible by 2, then we know that it's an even number. If a number is not divisible by 2, then we automatically know it's an odd number. In other words, it's an even number plus 1. So you can see how divisibility really allows us to see something about the number. In fact, one important means of detecting the personality of natural numbers is to factor them—that is, to write the number as a product of smaller numbers.

Those smaller natural numbers are called *factors*, and sometimes the factors of a number are also known as the *divisors* of the number. Let's see these ideas in action.

Let's suppose we consider 1 more than the product of two consecutive, even natural numbers. Our first reaction? Huh? Because there's a lot of stuff, a lot of ideas to parse. So let's try to ground our thinking—we always ground our thinking by looking at a few examples, and through the examples we might in fact see a pattern.

So, we're looking for two consecutive even numbers. That means two even numbers that are as close together as possible. The smallest such example would be 2 and 4. Those are two natural numbers that are even and as close as possible, because they're sandwiching the one number 3 in between. So, 2 and 4 are consecutive even numbers. We're to multiply them together, and then the rule we're told to do is to then add 1. So, we're looking at $2 \times 4 + 1$, which, notice, equals $8 + 1$, or 9.

Let's try another example. Let's take a look at the next pair of consecutive even numbers—they would be 4 and 6. So I would multiply 4×6 and then add 1. That would give us 24, plus 1 is 25.

If we take a look at one final example, let's take a look at the next pair of consecutive even numbers. That would be 6 and 8: $6 \times 8 + 1$ would be $48 + 1$, which equals 49.

Now let's look at the answers. We see 9, we see 25, we see 49. Well, what do we see? Surprisingly, we find structure. The answer always seems to be a perfect square. Notice that 9 is equal to 3 squared, 25 is equal to 5 squared, 49 is equal to 7 squared. It's almost like magic. In fact, the number that we square seems always to be the odd number right between the two consecutive even numbers with which we started: 3 is between 2 and 4, for example; 5 is between the 4 and 6; and 7 is between the 6 and 8. Remarkable.

This evidence leads us to create a conjecture, and so here's the conjecture we would create based on this evidence. Whenever 1 is added to the product of two consecutive even numbers, the answer is a perfect square. In fact, the answer is the square of the odd number between those two even numbers.

There's our conjecture, and now the question is, how can we prove that this pattern holds for all pairs of consecutive even numbers? It's not enough to just check a couple; we want to prove this is true in general.

We'll first inspire a way of thinking about the argument by considering the issue visually. Thinking about it visually is actually important, because it illustrates a wonderful life lesson that number theory teaches us. Whenever possible, we should try to look at issues from several points of view. Quite often, the vantage point we initially take on an issue might not be the one that's most efficient, most effective to allow us to build insight. Sometimes thinking about it differently allows us to actually see structure we otherwise would miss.

So, before going into the details of this mathematical rigorous proof, let's see why we should believe this remarkable theorem—remarkable conjecture; it's not a theorem yet—and see why this should make some sense.

I thought we'd try an experiment. We take a look at two consecutive even numbers visually. I'm going to look at 4 and 6. So 4 and 6. I'm going to represent the numbers by blocks, and this is not so easy to do, but look, I'm going to actually show it to you live. I've arranged them so you can see that what we have here is six blocks against the top, 1, 2, 3, 4, 5, 6; and then down here, I see 1, 2, 3, 4. So you could see that I have six blocks in this way going across and four blocks going down.

What we're supposed to do is consider the product of 4×6, and then we're supposed to add 1. Okay. So where is the 1? Well, here's the 1. Don't think I forgot. There's the 1 that we have to add. Somehow when we find $4 \times 6 + 1$, we're supposed to get a perfect square. What would that mean? Multiplying 4 by 6, you may recall, would be represented by looking at the area of this rectangle, or alternatively, how many blocks there are here. The number of blocks in this 4-by-6 rectangle actually equals the number 4×6.

How can I turn this configuration into a perfect square? Let's watch and see what happens if we think about this together. I'm going to remove this column right here. Now when I remove this column, what happens? I'm just going to lift it up so it's out of the way and we can look at it. Now I'm going to have five across since I removed one of the six: 1, 2, 3, 4, 5. But notice I still have four: 1, 2, 3, 4, down here.

I take this and I put it on its side, like so, and if I put it down, it fits right on top. So, now notice I have 1, 2, 3, 4, 5—one more than I had before, and here I have 1, 2, 3, 4, but there's a little hole—a hole for one cube. Notice that's the plus 1. When I put this in, almost like magic, what do I see? I see a 1, 2, 3, 4, 5 by 1, 2, 3, 4, 5—I see a 5-by-5 square. So therefore, how many cubes are there? Five times 5, or 5 squared: 25.

This idea actually is the idea behind the proof of the theorem in general. If I have one even number and then the next consecutive even number on top like this, if I were to always take up this last row and twist it up here like this, it would always fit on top with one extra space left over. I fill that space in, and I see a perfect square. This shows us that, in fact, visually, this idea makes some sense.

Let's see if we can now prove this in earnest. So, capturing this idea algebraically is what we now want to do.

We have two consecutive even numbers, but we don't know what those numbers are since we're trying to prove this for every single pair of consecutive even numbers. So we give them names. But we give them names where, in fact, those names could genuinely be anything. Here's the first place that we're going to introduce some unknowns. I'm going to call the first even number $2n$. So the n is any natural number at all. Notice that when I multiply by 2, it will definitely be an even number. That represents a generic even number.

If we have $2n$, what would be the next consecutive even number? I wouldn't add 1; I would have to add 2. I'd have to hop over that next odd number. So the next even number would be $2n + 2$. So, $2n$ and $2n + 2$ actually represent a generic pair of consecutive even numbers. We have to multiply them together, and then we have to add 1. First, let's multiply them together.

We can do that using the distributive property of multiplication, and I'll do it for you, so we're going to get a little technical here. We have $2n$ multiplied by the quantity $(2n + 2)$. When I distribute, I see $(2n \times 2n) + (2n \times 2)$. Well $2n \times 2n$ is $4n^2$, and $2n \times 2$ is $4n$. So I'm left with $4n^2 + 4n$.

To produce the number that we wish to study, we have to add on that 1, so I see $4n^2 + 4n + 1$. It's a mouthful, it's an algebraic mouthful, but we're done.

Now what do we hope that that's going to actually equal? We want to prove that this quantity equals the square of the odd number that's between those two evens. In other words, what's that odd number? It would be $2n + 1$, and I want to consider $(2n + 1)^2$. To square this quantity, we actually have to use the distributive property twice, so it's going to get a little bit tricky, but I'll do it for you.

We have the quantity $(2n + 1)$ multiplied by the quantity $(2n + 1)$. So when we distribute the first time, I see $2n + 1$ all multiplied by $2n$, plus $2n + 1$ multiplied by the number 1. Now we have to distribute yet again, and I would see $2n \times 2n + 2n$ plus another $2n + 1$. When we simplified the $2n \times 2n$, we see $4n^2$, and $2n + 2n$ is actually $4n$, and then I have a plus 1, which notice, is indeed equal to what we found previously.

So we see that 1 plus the sum of two even numbers, two consecutive even numbers, is exactly the square of the number that's in between. We've just shown that our conjecture holds in general. We proved it for every single pair of consecutive even natural numbers, since we never specified which ones we were looking at.

We could find a similar pattern with 1 plus the product of two consecutive odd numbers. For example, if we look at the first two odd numbers that are consecutive, 1 and 3, we multiply 1×3, we get 3, plus 1 is 4, and notice that's actually a perfect square, and it happens to be the perfect square of 2—2^2—which is between 1 and 3. So we seem to be seeing the exact same pattern, and I'll let you have the fun of making up your own conjecture and finding your own proof.

But congratulations to us for coming up with our first conjecture and actually devising our first proof.

There's another elementary question involving even and odd numbers that's easy to state, and I want to share it with you right now. It's one of the most famous questions from elementary number theory, and it's called the "$3n + 1$ question."

We begin with a process that generates a list of numbers, and to start we're given a natural number, and we'll call it n since we don't now what we're starting with and we don't want to specify it. Now we have two choices. If n is an even number, then the next number on our list that we're generating will be half of n, or $n/2$. If n is an odd number, then the next natural number on our list will be $3n + 1$. If n is even, then the next number will be smaller—half of n—if n is odd, then the next number will be larger. We multiply the number by 3 and add 1.

We now apply the same exact process with our new number to produce the next number on our list, and we repeat this process. This sequence was first studied by Lothar Collatz in 1937. As always, of course, such processes do not make any sense to us and really have no meaning until we consider examples. So let's consider some.

If we start with a 1 as our first number, then what do we see? Well, 1 is an odd number, so what's the rule? We multiply by 3 and add 1 to get the next one: 3×1 is 3, plus 1 is 4. So our next number is 4. So we see a 1 and then a 4.

That number 4 is actually even, so that means we take half of it. The next number would be then half of that, which is 2. Two still is even, so we take half of it, and we see 1. Well, 1 we've already seen. We know what to do: We multiply it by 3 and add 1, and we get 4. Four, we take half of it, we get 2, take half of it, we get 1. The pattern is clear. In fact, let's notice that a more general property actually holds.

If at any point we generate the number 1, then we know that the rest of the list will just repeat: 1, 4, 2, 1, 4, 2, 1, 4, 2, and so on, forever.

What if we start with a 2? Then we see that if we start with a 2, it's even, so I take half of it—that gives me a 1. One is odd, so I multiply it by 3 and add 1, that gives me 4, and I'm right back where I started from. Again, we see that this list is going to end with 1, 4, 2, 1, 4, 2, and so forth, forever.

What if we start with a 3? This is genuinely going to be new. Start with a 3, that's odd, so the next number will be $3 \times 3 + 1$, which is 10. Ten is even, so I take half of it to get the next number, which would be 5. Five is odd, so I multiply by 3 and add 1, that's $15 + 1$, which is 16. Sixteen is even. Take half of it: 8. Eight is even. Take half of it: 4. Four is even. Take half of it: 2. Two is even. Take half of it: 1. Once we hit 1, we know, we've already seen, we get a 4, a 2, a 1: 4, 2, 1, and so forth, forever.

Interesting. Let's start with a 4. Well, that's actually easy. If we start with a 4, we take half of it and get 2, take half of 2, we get 1, and then automatically we cycle back: 4, 2, 1, 4, 2, 1, 4, 2, 1, forever. Are you seeing the pattern?

Let's try one last example. Let's start with 11, really different. Now 11 is odd, so we start off by multiplying by 3—that's 33—and adding 1—that's 34. Thirty-four is even, so I take half of it, which is 17. Seventeen is odd, so I multiply it by 3 and add 1 and get 52. Fifty-two is even, so I take half of it. The next number is 26. Twenty-six is still even, so I take half of it and I get 13. Now 13 is odd, so I multiply that by 3, get 39, and add 1, which is 40. Forty is even. The next number would be half of it: 20. Twenty is even. Half of it: 10. Ten is even. Half of it: 5. Now, 5 we've actually already seen. What do we do? After 5 comes 16, then 8, 4, 2, 1, 4, 2, 1, 4, 2, 1, and so forth.

Yet again we see a tail of an endless repeated run of 1, 4, 2, 1, 4, 2. Absolutely amazing. Notice how the numbers in the list go up and down in size. In fact, sometimes these sequences are actually called the "hailstone sequences," because as many of you might know, hailstones are formed in high clouds when a small particle gets coated with layers of ice and then moves up and down in the high clouds until it becomes so heavy that it actually falls to Earth. We're seeing that kind of pattern here wiggling.

Our experimentation seems to be leading us to a conjecture. It appears that no matter what natural number we start with, the list always eventually becomes the endless run of 1, 4, 2, 1, 4, 2, and so forth.

If we start with a relatively modest number, such as 27—and you can try this if you want, but I don't actually suggest it—the process will produce a list of 111 numbers before we finally come to our first 1, and within those first 111 numbers, the numbers we'd see would actually go as large as 9,232—so, huge. But, we do finally settle down to the familiar 1, 4, 2, 1, 4, 2, and so forth.

Well, mathematicians and computer scientists have actually tested every natural number up to around 3×10^{18}—that's the number 3 followed by 18 zeros. When you start with all of these values up to that point as a starting value, in each case, the list finally does settle down into the familiar 1, 4, 2, 1, 4, 2, and so forth.

Collatz himself conjectured that starting with *any* natural number, the list will eventually settle down to the familiar 1, 4, 2, 1, 4, 2 that we've seen. But no one has been able to prove this conjecture in general. The Collatz conjecture remains one of the most famous open questions in elementary number theory.

The great 20th-century Hungarian mathematician Paul Erdös, who we'll talk about in Lecture Nine, actually offered a $500 prize for solving this problem, and he once wrote, "Mathematics is not ready for such problems," meaning that we don't even have the machinery to allow us to make sense of these things. So questions even that sound simple remain notoriously hard and difficult.

I want to close our lecture here by considering one last observation about the most even numbers around, the powers of 2. Let's just produce the first few powers of 2: 2 to the first power is just 2, 2^2 is 4, 2^3 is 8, 2^4 is 16, 2^5 is 32, 2^6 is 64. If you go out a little bit, 2^9 is 512, and if you really zip up, you can get to 2^{24}, which is 16,777,216.

As you look at these screeds of digits that we have here, making up the powers, I want us to consider the following question: What are the first—meaning the left-most—digits that we see in the previous numbers? What are those first digits in the powers of 2? We saw a 2, 4, 8, a 1, a 3, a 6, and then later, a 5. The only digits missing from our list, in fact, were just 7 and 9, kind of interesting.

Are there powers of 2 that begin with the missing nonzero digits 7 and 9? Using a computer, one can verify that 2^{46} equals 70,368,744,177,664, but it starts with a 7; and 2^{53} actually is a 16-digit number that begins 9,007..., and so on.

What we see here is that there are powers that begin with every single digit from 1 to 9 appearing in the very first, left-most spot of a power of 2. Is there a power of 2 that begins with a 10, or with an 11, or with a 12? Is there a power of 2 that begins with your social security number? How about your social security number, immediately followed by your birth date, followed by your cell phone number including its area code? That's a really big natural number.

Here's the surprising fact. As amazing as it might first seem, there is a theorem that asserts that, given any natural number n at all, there exists a power of 2 whose left-most digits agree with the digits of the given natural number n. The truly amazing aspect of this assertion is that this result can be proved for all natural numbers.

The proof of this theorem requires many different ideas from number theory that we ourselves will explore throughout our journey together. In fact, to prove the result, we must actually leave the comfortable world of the natural numbers and apply results involving irrational numbers, which we haven't even talked about yet.

This connection will be our final illustration of the incredible synergy between the many different types of numbers. So for now, I leave this whole realm of elementary questions, and I leave with the reasoning that establishes this theorem as a number theory cliffhanger that will be resolved at the very close of this course.

Lecture Three
Triangular Numbers and Their Progressions

Scope:

On the one hand, studying Pythagoras, the celebrated ancient Greek lover of number from the 6th century B.C.E., is a natural beginning to this lecture. On the other hand, it seems fitting to open with the great 19th-century German mathematician Carl Friedrich Gauss, who has contributed either directly or indirectly to nearly every advance of every corner of number theory and is known as the "Prince of Mathematics." Here we will bridge both ages and cultures to study a common mathematical curiosity that each of these two prominent figures explored—the collection of figurate numbers known as *triangular numbers*. Triangular numbers can be defined intuitively as the number of billiard balls required to create larger and larger equilateral triangles. As we will see, this basic idea leads to some very profound theorems about the natural numbers. This discussion will foreshadow our later discussion on searching for natural-number solutions to certain equations—an area now known as *Diophantine analysis*. The triangular numbers have important implications within both our everyday world and the world of number theory. Generalizing these numbers will allow us to discover the central mathematical concept of arithmetic progressions. Arithmetic progressions are lists of numbers for which the difference between any two adjacent terms is always the same value. Both the even numbers (2, 4, 6, 8, and so forth) and the odd numbers (1, 3, 5, 7, and so forth) are examples of arithmetic progressions. These arithmetic progressions of numbers are central and useful objects within the study of numbers and will follow us throughout our studies.

Outline

I. The mathematical life and mind of Carl Friedrich Gauss.

    A. Gauss's early days.

        1. Carl Friedrich Gauss was born on April 30, 1777, in Germany. His father was a bricklayer and did not encourage his son to pursue advanced mathematics.

2. However, it was clear from a very early age that Gauss was a true mathematical child prodigy. There are many stories about young Gauss.
3. One famous story involves Gauss as a third-grade student and his teacher J. G. Büttner. Gauss discovered a very clever proof of the formula: $1 + 2 + 3 + \cdots + n = n(n + 1)/2$.

B. The "Prince of Mathematics."
1. Today Gauss is considered one of the greatest mathematicians ever.
2. Gauss's passion toward number theory was clear. He believed that basic relationships between numbers were fundamental to all matter. He once wrote that "God does arithmetic," and is credited with having said, "Mathematics is the queen of the sciences, and number theory is the queen of mathematics."
3. Gauss was later crowned the "Prince of Mathematics."
4. He was a perfectionist, and thus his accomplishments were much more numerous than his publications. One of his mottos was: "Few, but ripe."
5. Another favorite motto of Gauss's came from Shakespeare's *King Lear*: "Thou, nature, art my goddess; to thy laws my services are bound ... ."

II. Triangular numbers.

A. A geometric pattern of numbers.
1. The sum of the first *n* natural numbers is called the nth *triangular number.*
2. The first few triangular numbers are 1, 3, 6, 10, 15, 21, and 28.
3. These numbers are called *triangular* because they represent the number of billiard balls that can be arranged into an equilateral triangle. Notice that the fourth triangular number, 10, revered by the Pythagoreans, equals the number of billiard balls used in a game of pool or the number of pins used in bowling.
4. Applying Gauss's elementary school formula, we see that 5050 balls are required to make a triangle with a base of 100 balls.
5. We can use the triangles to find the sum formula.

B. Turning triangles into squares.

    1. In an attempt to find structure within these numbers, we consider the sums of two consecutive triangular numbers.

    2. The first few of those sums equals 4, 9, 16, 25, 36, and 49. We cannot help but see a surprising pattern—the sums are perfect squares.

    3. We can establish that this observation is a theorem that holds in general.

    4. This result can be proven algebraically or geometrically.

C. Gauss's great theorem.

    1. On July 10, 1796, at the age of 19, Gauss proved that every natural number is the sum of at most three triangular numbers. For example: $17 = 1 + 6 + 10$, and $100 = 45 + 55$.

    2. He stated this result in his diary as "Eureka! Num $= \Delta + \Delta + \Delta$." In fact, Pierre de Fermat asserted this result back in 1638. He claimed to have a proof of this result, but no such proof has ever been found.

    3. Establishing this result is an extremely challenging proposition. Given a natural number N, we must find whole numbers x, y, and z that satisfy the Diophantine equation $x^2 + y^2 + z^2 + x + y + z = 2N$, and this foreshadows our explorations into finding solutions to such equations later in our course.

III. A "shaky" application.

  A. A hand-shaking question.

    1. Suppose that a group of people assemble for a meeting. Before the proceedings begin, each pair of people shakes hands exactly once. How many handshakes are there?

    2. With two people we have 1 handshake; with three people we have 3 shakes; with four people we have 6 shakes; with five people we have 10 shakes.

    3. We discover that the number of handshakes will be a triangular number.

    4. While this application might appear frivolous, such counting issues have applications in networking. If n computers are to be directly connected with each other, then the number of connections would be exactly the nth triangular number.

B. A sample of combinatorial number theory. Counting complex scenarios is challenging, and that area of mathematics is known as *combinatorics*. The handshake question is an example of what is known as *combinatorial number theory*.

IV. The notion of arithmetic progressions.

A. Extending triangular numbers.

1. The triangular numbers arose from adding the numbers from the very simple progression 1, 2, 3, 4, 5, and so forth.

2. What feature makes the progression of natural numbers so simple? To generate the next number on our list, we merely add 1 to the previous number.

3. We can now generalize this feature. Consider the progression that starts with 1, and to generate the next number we add 2 to the previous number. This new progression is: 1, 3, 5, 7, 9, 11, 13, and so forth. We have produced the odd numbers.

B. An introduction to arithmetic progressions.

1. Progressions of numbers for which the next number in the list is produced by adding a fixed amount to the previous number are called *arithmetic progressions*.

2. So the lists 1, 2, 3, 4, … and 1, 3, 5, 7, 9, … are both examples of arithmetic progressions. In fact, the even numbers—2, 4, 6, 8, 10, … —also form an arithmetic progression.

3. We can build any arithmetic progression just by knowing the first number together with the fixed amount to be added to create each successive number. For example, if we start with 5 and add 3 each time, then we have the arithmetic progression 5, 8, 11, 14, 17, 20, 23, 26, 29, 32, and so forth.

C. A formula for the sum of the terms in an arithmetic progression.

1. The triangular numbers arose from adding up the first few numbers from the simple arithmetic progression 1, 2, 3, 4, and so forth.

2. We saw that $1 + 2 + 3 + \cdots + n = n(n + 1)/2$.

3. So one generalization to the triangular numbers can be found by considering the sums of the first few terms from any arithmetic progression. For example, if we return to the arithmetic progression 5, 8, 11, 14, 17, 20, 23, 26, 29, 32, and so forth, then we can generate the list of sums of these numbers. That list would begin 5, 13, 24, 38, 55, 75, 98, and would provide one type of generalization of the triangular numbers.

4. Beyond extending the notion of triangular numbers, arithmetic progressions are one of the pillars of number theory and will prove to be extremely useful in our future explorations.

Questions to Consider:

1. Compute the sum of the first 1 million natural numbers.

2. Consider two consecutive triangular numbers, such as 3 and 6. Square each number and then subtract: $36 - 9 = 27$. Perform this with several examples. What do you notice about your answers? Make your own conjecture. (*Bonus*: Can you prove your conjecture holds in general?)

Lecture Three—Transcript
Triangular Numbers and Their Progressions

There are many ways to begin in earnest our adventure into numbers and their arithmetic. On the one hand, a natural place from which to begin our journey is with one of the forefathers of the subject— Pythagoras, the celebrated ancient Greek lover of numbers from the 6th century B.C.E.

On the other hand, it seems fitting to open with the great 19th-century German mathematician Carl Friedrich Gauss, who has contributed either directly or indirectly to nearly every advance, of every corner, of number theory and now is known as the "Prince of Mathematics."

Here we'll bridge the ages to study a common mathematical curiosity that each of these two prominent figures explored—the collection of figurate numbers known as *triangular numbers*.

Triangular numbers can be defined intuitively as the number of billiard balls required to create larger and larger equilateral triangles. As we'll see, this basic idea leads to some very profound theorems about the natural numbers. In fact, this discussion will foreshadow our later discussion on searching for natural-number solutions to certain equations—an area now known as *Diophantine analysis*.

The triangular numbers have important implications both within our everyday world and the world of number theory. We'll see, for example, that the triangular numbers model the number of hands that would be shaken within any friendly encounter among a group of individuals. Generalizing these numbers will allow us to discover the central mathematical concept of arithmetic progressions. Arithmetic progressions are lists of numbers for which the difference between any two adjacent terms is always the same value. Both the even numbers—2, 4, 6, 8, and so on—and even the odd numbers—1, 3, 5, 7, and so forth—are examples of arithmetic progressions, because in each case any two adjacent terms differ by 2. These arithmetic progressions of numbers are central and important objects within the study of numbers and will remain with us throughout our studies.

I want to open this lecture with a few words about the mathematical life of Carl Friedrich Gauss. Gauss was born on April 30, 1777, in Germany. His father was a bricklayer and actually did not encourage his son to pursue advanced mathematics. However, from a very early age, it was clear that Gauss was a true mathematical child prodigy. There are many wonderful stories about a very mathematically precocious young Gauss. One famous story—which, by the way, may or may not be true, but still remains a wonderful story— involves Gauss as a third-grade student and his teacher at the time, a J. G. Büttner. Büttner one day asked his students to add up the first 100 natural numbers, thinking that this would keep the class busy for some time. Perhaps Büttner didn't bother to actually prepare a lecture for the day. Who knows?

Anyway, within a few minutes, Gauss approached the front of the class with a piece of paper in his hand. Gauss asserted that the answer to the sum was 5050, which, by the way, is the correct answer. In the story, Büttner asks Gauss how he computed this sum of the first 100 numbers so quickly, so Gauss describes his clever idea, which I want to share with you now.

What Gauss does is a lesson that we've seen in the previous lecture, which is to look at things from different points of view. We take the sum, which, let's just call it s, to give it a name, since we don't know its value yet. So s will equal $1 + 2 + 3 + \cdots$, and so forth, all the way out to $+ 99 + 100$. Now Gauss does something quite ingenious. He writes that exact same sum again but with the terms written in reverse order. So he writes s again as $s = 100 + 99 + 98 + \cdots$, and so forth, all the way down to $+ 2 + 1$.

Then he adds these two equalities. On the left-hand side, we see $s + s$ which gives us $2s$. On the right-hand side, notice that the columns of numbers always sum to the same value: $1 + 100$, in the first column, equals 101; $2 + 99 = 101$; and all the way down to the penultimate one—$99 + 2$, which is 101—and even the last one, $100 + 1$ is 101. So we see 101 added to itself a whole bunch of times, and how many? Well, one for each natural number from 1 to 100. Therefore, we see 100 times 101.

Gauss produces a wonderful equation, which is that 2 times the sum—$2 \times s$—equals 100 multiplied by 101. We can solve for s by itself by dividing both sides by the 2. So we see s equals 100 times by 101, all divided by 2, which equals 5050. Wonderful.

In fact, Gauss's ingenious idea can be generalized. Let's find the sum of the first n natural numbers where we don't even know the actual precise value for n. We'll do this now in general, and we'll use exactly Gauss's ideas, we'll follow Gauss's footsteps—his young footsteps.

Let s equal the sum of the first n natural numbers, so $s = 1 + 2 + 3$, all the way out to $(n - 1)$, plus the last one, n. Now we write the exact same sum again but reverse the order of the summands. So $s = n$, plus $(n - 1)$, plus, and so forth, decreasing each time until we get down to $+ 2 + 1$.

When we add both of these together, we see $s + s$ is still $2s$, and now each column adds to $n + 1$. One $+ n$, in the first column, is $n + 1$; $2 + (n - 1)$ is $n + 1$, and so forth, all the way down to the last term, $n + 1$, which is $n + 1$.

How many $(n + 1)$s are we adding together? One for each natural number from 1 to n. So there's n of them, and so we see n multiplied by $(n + 1)$.

If we divide both sides by 2, we come up with a beautiful formula: $s = n(n + 1)/2$. Therefore, Gauss discovered a very clever proof of this formula: $1 + 2 + 3$, all the way out to $+ n$, equals $n(n + 1)/2$.

This formula is really quite a work of beauty, and it transforms a seemingly intractable sum to a very manageable form, which we call a "closed formula"—a formula that has no "dot, dot, dots" in it but just is totally compact. For example, if we want to find the sum of the first 1,000,000 natural numbers, we now see that the answer is 1,000,001 × 500,000. Not bad for a third grader, don't you think? Pretty clever guy.

Today, Gauss is considered one of the greatest mathematicians ever. Gauss's passion toward number theory was clear. In fact, he believed that basic relationships between numbers were fundamental in all matter, even beyond mathematics. He once wrote that, "God does arithmetic," and is credited with having said that, "Mathematics is the queen of the sciences, and number theory is the queen of mathematics." So you could really see Gauss's passion not only for math in general, but number theory in particular. Later, Gauss was actually crowned the "Prince of Mathematics" by the mathematical community, and that is his moniker to this very day.

He was a perfectionist; thus his accomplishments were much more numerous than his actual publications. In fact, one of his favorite mottos was the phrase "Few, but ripe," meaning he wouldn't publish lots and lots of little results, but instead the results that he did publish were quite important and quite dramatic. Another favorite motto of Gauss's, by the way, comes from Shakespeare's *King Lear*, in which we see, "Thou, nature, art my goddess; to thy laws my services are bound …" So you can see Gauss's connection between nature, mathematics, and numbers.

Gauss is so respected and admired that his image actually appeared on the 10 deutsche mark notes before the euro came into fashion. Here you can actually see Gauss on real money. Not bad.

Let's now return to the sum of the first n natural numbers, which we found equals $n(n + 1)$, all divided by 2. This sum is called the nth *triangular number*. The first few triangular numbers, therefore, if we just plug in, we see 1, 3, 6, 10, 15, 21, and so forth, and we can see this by just computing. We have 1, that's just 1. The next one would be $1 + 2$, which would be 3, the next number on this list: $1 + 2 + 3$ gives us 6. If we add 4, we get 10, and so forth. So we can see that these numbers are correct.

These numbers are called *triangular* because they represent the number of billiard balls that can be arranged into an equilateral triangle. Notice that the fourth triangular number—10—equals the number of billiard balls used in the game of pool, or the number of pins used in bowling. By the way, notice that the deutsche mark that has Gauss's picture on it is a 10 deutsche mark—10, triangular number. Coincidence? Who knows.

Anyway, applying Gauss's third-grade formula, we see that 5050 balls are required to make a triangle having a base of 100 balls. So, an equilateral triangle that would be 100 here, 100 here, and 100 here would require 5050 balls.

We can use the visual description of these numbers as triangles to actually give a geometric proof of Gauss's sum formula, and I want to share that with you right now.

If we take a look at the triangular number 10, which is the familiar number that we see if we play billiards or if we bowl, you can see that it forms an equilateral triangle. We have $4 + 3 + 2 + 1$, and that gives us the 10. Wonderful. Now I want us to rearrange this triangle a little bit, no longer have it be equilateral, but I'm going to create a right triangle, by just slightly sliding over these balls.

I want you to notice something. This is almost like a magic trick. There's no sleight of hand; I'm never actually removing any of these billiard balls, but instead, I'm just rearranging them a little bit by sliding them over. Notice that in doing so, I actually create a right triangle, but the number hasn't changed. It's still the triangular number 10, and you can count and see 4 by 4.

This lends itself to some wonderful observations, because to prove the formula of Gauss I take another copy of this triangular number. Notice that here I have another copy, which you can see is indeed also 4 by 4. You can count: 1, 2, 3, 4, 5, 6, 7, 8, 9, 10. Great. So these are identical. Look what happens when I turn this around: The two fit together, and I form a rectangular figure. Do you see it? Amazing.

If I can now find the area of this, which is very easy to do, what's the area? It's 1, 2, 3, 4, multiplied by 1, 2, 3, 4, and then an extra 1 here, so it's 5. It's 4×5, which, of course, is 20. There's 20 dots here. But remember, I've overcounted because in fact I added a whole extra copy, so I actually have double the actual answer. If I were to divide by 2, then what I would see would be 20 divided by 2, or 10, and that's exactly how many dots we have that are orange.

Let's say this again. By taking an extra copy—so now I have 2 times the number of dots—turning them around, and placing them in this configuration, I see a rectangle, and the rectangle's base is $4 + 1$, and the height is 4. Therefore the area will be $4 \times (4 + 1)$. I divide by 2, and that gives us the number of dots here.

In fact, this geometric proof holds for any number of dots, as long as they form a triangular number. Think about it, if this were n by n, then when I put in this extra copy, I would see n multiplied by $(n + 1)$. So I'd see $n(n + 1)$, but I overcounted. I doubled the answer, so I have to divide by 2. When I divide by 2, I come up with the formula that we actually saw in Gauss's formula: $n(n + 1)$, all divided by 2.

In an attempt to find structure within these numbers, we consider the sums of two consecutive triangular numbers. Again, always searching for structure, always searching for a pattern. Let's take two consecutive numbers in this triangular sequence and add them together. Here are the first few: We have $1 + 3$, which equals 4; $3 + 6$ equals 9; $6 + 10$ equals 16; $10 + 15$ equals 25; $15 + 21$ equals 36; $21 + 28$ equals 49.

We can't help but see a surprising pattern. The sums are perfect squares: 4, (2^2); 9, (3^2); 16, (4^2); 25, (5^2); and so on. We can actually establish that this observation holds in general, and is, in fact, a theorem. We can see why this result holds by returning to our visualization of these numbers as triangles. Let's first try to do that right now live and then see the actual proof.

Here, this is the fourth triangular number. What if I add to it the third triangular number, which by the way, is 6? I happen to have the third triangular number in its configuration right here. You could see it's 3. If you add the dots up, we see, 1, 2, 3, 4, 5, 6. So that's right. What if we add these dots to these dots? Well, doing that like this doesn't seem to really reveal anything, but watch the magic when I turn this around and fit it in place.

Isn't that amazing? Like magic, we form a perfect square. It's a 1, 2, 3, 4 by 1, 2, 3, 4—it's a 4-by-4 square. So the number of dots in the sum of the orange plus the white is a perfect square number, in this case, 16. Really, really neat. That shows us that this makes some sense, but now we can actually see why this is true for real.

This result can also be proved using algebra, and so let's take a moment to think about this. If we add the nth triangular number to the $(n + 1)^{th}$ triangular number, then what would we see? We're going to do some algebra here, so let's just have some fun and relax.

The nth triangular number by Gauss's formula is $n(n + 1)$, all divided by 2. Now I'm going to add the next triangular number, which means that we have to add 1 to all those little n's. So I see $(n + 1)(n + 2)/2$. That's the next triangular number. What do we do here? We just combine those two fractions, and so I see $(n(n + 1) + (n + 1)(n + 2))/2$.

If you notice, in the numerator, we have this common factor of $(n + 1)$ in both the first term and the second term. So I could actually factor that out, and when I factor it out, I'm left with $(n + 1)$ times the quantity $(n + n + 2)$, and that's all divided by 2. What's $(n + n + 2)$? That's $(2n + 2)$.

So I see $(n + 1)(2n + 2)/2$. Notice that in the $(2n + 2)$ factor, I can actually factor out another number, namely, 2. When I factor out that 2, I see $2(n + 1)(n + 1)/2$. The 2 divided by 2 cancel, and I'm left with $(n + 1)(n + 1)$, which is $(n + 1)^2$.

This establishes the result in general, since I never said what n was. We see that whenever we add two consecutive triangular numbers, their sum will always be a perfect square.

On July 10, 1796, at the age of just 19, Gauss proved that every natural number is the sum of at most three triangular numbers. Absolutely amazing. For example, if we take a look at the number 17, that equals $1 + 6 + 10$, each of which we've seen is a triangular number. Even a number like 100 equals $45 + 55$, and you can check that both 45 and 55 are, indeed, triangular numbers.

He stated this result in his diary as, "Eureka! Num = $\Delta + \Delta + \Delta$." He actually drew little triangle, plus triangle, plus triangle. He was really delighted to discover a proof of this really deep mathematical result.

Pierre de Fermat asserted that this result holds way back in 1638. Fermat claimed to have a proof of this result, but no such proof has ever been found. We'll talk about Fermat quite often in this course, and we'll see this is a recurring theme. Fermat states assertions and claims to have proofs, and often does not have a proof, or certainly the proof wouldn't be necessarily complete. It would be correct, but not necessarily complete.

Establishing this result is extremely challenging—a very difficult proposition, indeed. Let's just think about what would be involved. Given a natural number, let's call it N, we would have to find whole numbers—so natural numbers or maybe 0—x, y, and z that satisfies the Diophantine equation: $x(x + 1)/2 + y(y + 1)/2 + z(z + 1)/2 = N$. Now where did that come from?

Think about it: Each one of those represents a triangular number. That's the formula for the triangular numbers. So it's saying that one triangular number formed by the x's, plus another triangular number formed by the y's, plus the third triangular number formed by the z's has to equal the given number N. Now if we actually multiplied everything through by 2, the denominators on the left-hand side would go away, and we'd see a 2 on the right-hand side multiplying N. If we distributed out all those multiplications so there's no more parentheses, this would become the equivalent equation $x^2 + y^2 + z^2 + x + y + z = 2N$.

We have to find whole-number solutions x, y, and z that satisfy this very complicated equation for any natural number N. Very tricky, indeed, although this issue does foreshadow our explorations into finding solutions to such equations later in our course. So we'll come back to this, although we won't solve this question since it really is genuinely difficult, indeed.

I wanted to take a moment to consider an application in which these triangular numbers actually arise. Let's suppose that a group of people assemble for a meeting. Before the proceedings begin, each pair of people shakes hands exactly once. How many handshakes are there? Let's think about that.

With two people, there's, of course, just 1 handshake—there's two people; they shake hands with each other. With three people, we have that 1 handshake, but now this new person has to shake hands with the two previous people, and so we see now 2 more handshakes, so a total of 3 shakes.

Now we have 3 handshakes. If we introduce a fourth person, that person has to shake hands with all three people, which means that now we have 6 handshakes. With five people, we would have 10 handshakes, and so forth. We discover that the number of handshakes will be a triangular number. Who would have guessed it? Something from number theory actually helps us count something. It's rare, but it's wonderful when it happens.

While this application might seem a little bit frivolous, such counting issues have applications in networking. If n computers are to be directly connected with each other, then the number of connections would, in fact, exactly be the nth triangular number.

Counting complex scenarios such as handshakes or computer networks is challenging, and that area of mathematics is known as *combinatorics*. The handshake question is an example of what is known as *combinatorial number theory*.

Let's now generalize the idea of triangular numbers in a slightly different direction. Recall that the triangular numbers arose from adding the numbers from a very simple progression: 1, 2, 3, 4, 5, and so on. What feature makes the progression of natural numbers so simple? Well, to generate the next number on our list, we merely add 1 to the previous number. We can now generalize this feature. So we pull out this feature and see if we can extend it.

Let's consider the progression that starts with 1, and to generate the next number, we add 2 to the previous number. This new progression is 1, 3, 5, 7, 9, 11, 13, and so forth. We've just produced the odd numbers.

Progressions of numbers for which the next number in the list is produced by adding a fixed amount to the previous number are called *arithmetic progressions*. We're performing addition—arithmetic—to progress the numbers. So the list 1, 2, 3, 4, 5, and so forth, and even the odd numbers 1, 3, 5, 7, 9, and so forth, are both examples of arithmetic progressions, because in each case there's a fixed number that I can add to one number to get the next one.

In fact, the even numbers: 2, 4, 6, 8, 10, and so forth, also form an arithmetic progression, because if I start with 2 and then continually add 2, that gives me 4, add 2 again gives me 6, add 2 again gives me 8, and we see this progression.

We can build any progression just by knowing the very first number together with the fixed amount to be added to create each successive number. For example, if we start with a 5 and add 3 each time, then we have the arithmetic progression: 5—now to get the next one, we have to add 3, so that gives us 8. Now we take 8 and we add 3 yet again to get 11. So we see the arithmetic progression: 5, 8, 11, 14, 17, 20, 23, 26, 29, and so forth.

We discovered the triangular numbers by adding up the first few numbers from the simple arithmetic progression 1, 2, 3, 4, 5, and so on. Recall that we found that the formula $1 + 2 + 3 + 4$, all the way out to $+ n$, equals $n(n + 1)/2$.

So, one generalization of triangular numbers arises by considering the sums of the first few terms from any arithmetic progression. For example, if we return to the arithmetic progression we just saw—5, 8, 11, 14, 17, and so forth, we keep adding 3 each time—then we can generate the list of sums of these numbers as we progressively add the next term. That list would begin 5—the first number—then $5 + 8$, which is 13; then $5 + 8 + 11$, which is 24; and then $5 + 8 + 11$ plus the next one, which is 14, which equals 38; and so forth. So our sequence of sums would be 5, 13, 24, 38, 55, and so forth, and would provide one type of generalization of the triangular numbers.

Beyond extending the notion of triangular numbers, arithmetic progressions are one of the pillars of number theory and will prove to be extremely important in our future lectures.

So, while we've seen the wonderful world of arithmetic progressions where we have this very, very simple pattern—and how in the very simple case where we just look at the arithmetic progression of the natural numbers themselves, we come to these wonderful triangular numbers—we will take the notion of arithmetic progression with us, and we'll actually see it as a recurring theme throughout our studies.

For now, I want you to enjoy the triangular numbers, and not only the wonderful theorems that we discovered about the structure of these numbers, but perhaps more importantly, the beauty of the proofs, both geometric and algebraic, that proved the assertions were in fact a fact.

Lecture Four
Geometric Progressions, Exponential Growth

Scope:

Elementary number theory is not a euphemism for "easy number theory." Instead it is an area of number theory that has its focus on fundamental questions about numbers, most of whose subtle answers do not involve advanced mathematical techniques. Here, with the notion of an arithmetic progression—a list of numbers generated by successive addition—fresh in our minds, we open these lectures on elementary number theory by exploring the multiplicative cousin of arithmetic progressions. These are known as *geometric progressions*— number lists generated by successive multiplication. These numbers grow at a dramatically fast rate and possess enormous structure. We will momentarily pause to consider how geometric progressions naturally appear in the world of music as a means of producing an even-tempered scale. Next we will turn to an extremely important and useful object in the study of advanced number theory: the *sum* of the terms of geometric progressions. Our exploration into these sums of numbers will include some amusing ancient stories that hold important mathematical morals. We will then extend this additive issue and consider the *endless* sum of *all* terms of certain geometric progressions and make such vexing ideas intuitive by visualizing such sums geometrically. The endless sum of a geometric progression is known as a *geometric series*. These series are fundamental in all corners of mathematics and science, especially—as we will see in future lectures—in number theory itself.

Outline

I. Geometric progressions.

   A. Introducing the notion of geometric progressions.

      1. An arithmetic progression is a list of numbers with the property that to get from any number to the next we need only add a fixed number. As we will discover as our course unfolds, these arithmetic progressions, while simple in structure, play an important role in our number theory story.

2. Perhaps even more important are the corresponding progressions in which the addition is replaced by multiplication. These progressions are called *geometric progressions*.

   3. To generate a geometric progression, we must be given the first number and then the constant multiple (known as the *ratio*) that generates successive numbers on the list.

B. Examples and illustrations.

   1. For example, if we start with 1 and are given the ratio of 2, then our geometric progression equals: 1, 2, 4, 8, 16, 32, 64, and so forth. It is easy to see that a generic term in this geometric progression is given by 2^n for some natural number n.

   2. The constant multiple (in the previous example, 2) is called the *ratio* because the ratio of any two consecutive terms—the larger to the smaller—always equals that fixed ratio. For example, notice that $32/16 = 2$.

   3. In general, if we start with 1 and have a ratio of r, then we would generate the geometric progression 1, r, r^2, r^3, r^4, r^5, and so forth.

C. A formula for the terms and exponential growth.

   1. These examples show us that a general term in a geometric progression is of the form r^n, for some whole number n.

   2. Since the varying quantity is the exponent n, we say that r^n *grows exponentially* if $r > 1$; and for $0 < r < 1$, we say r^n *decays exponentially*.

   3. If we start with 1 and consider $r = 1/2$, then we have a geometric progression that decays exponentially: 1, 1/2, 1/4, 1/8, 1/16, 1/32, … .

II. Progressions in music.

A. Ratios of pitches.

   1. A musical interval is an *octave* if the two pitches have frequencies in a ratio of 2:1. An interval is a perfect fifth if the ratio of the frequencies of the two pitches is 3:2.

   2. For example, in modern Western music, the A above middle C has a frequency of 440 Hz. Thus, the A one octave higher has a frequency of $440 \times 2 = 880$ Hz, and the E one-fifth higher than A440 has a frequency of $440 \times 3/2 = 660$ Hz.

B. The chromatic scale.

    1. In Western music, the *chromatic scale* begins at one pitch, say A, and progresses up in what are called *half steps* until it ends with the note that is one octave above the starting pitch. The pitches are derived from a progression of perfect fifths, starting with the first pitch.

    2. A progression of perfect fifths is a geometric progression with $r = 3/2$. If we start at A (440 Hz), we would produce: 440, $440 \times 3/2$, $440 \times (3/2)^2$, $440 \times (3/2)^3$, $440 \times (3/2)^4$, $440 \times (3/2)^5$, and so forth; that is: 440, 660, 990, 1485, 2227.5, 3341.25, and so forth.

    3. In order to keep the notes within the 440–880 Hz range, we divide the frequencies by 2 in order to lower the pitches so that they fall into the correct octave. This process requires us to modify our attractive geometric sequence.

    4. So why do we end up with a 12-note chromatic scale? The answer to this question—which we will discover for ourselves later in the course—involves irrational numbers, and we leave it as a musical and mathematical cliffhanger for now.

III. Summing geometric progressions.

  A. Seeking a pattern within a sum.

    1. It will be extremely useful to find the sum of the first terms of a geometric progression, just as we saw with arithmetic progressions.

    2. If we consider the geometric progression 1, 3, 9, 27, 81, … , then the list of the first five successive sums is: 1, 4, 13, 40, 121. A pattern is not immediately apparent.

    3. To build intuition, we will explore a particular case with some care. If we let $S = 1 + 3 + 3^2$ (which equals 13), then $3S = 3 + 3^2 + 3^3$. Subtracting, we discover:

$$\begin{array}{r} 3S = 3 + 3^2 + 3^3 \\ - \ \underline{S = 1 + 3 + 3^2} \\ (3-1)S = -1 + 3^3. \end{array}$$

    4. So we find that $S = (3^3 - 1)/(3 - 1)$.

5. We can see this pattern holds for other sums. For example, applying the analogous pattern with $1 + 3 + 3^2 + 3^3 + 3^4$, we would conjecture that the sum equals $(3^5 - 1)/(3 - 1)$, which equals $(243 - 1)/2$, which equals 121, as expected.

B. Finding a formula.

    1. We now generalize our example for finding the sum $1 + r + r^2 + r^3 + \cdots + r^n$, for any ratio $r \neq 1$. We call this sum S.

    2. If we multiply S by r, we can align most of the terms in S with most of the terms in rS and then subtract:

$$rS = r + r^2 + r^3 + \cdots + r^n + r^{n+1}$$
$$- \underline{S = 1 + r + r^2 + r^3 + \cdots + r^n}$$
$$(r - 1)S = -1 + r^{n+1}.$$

Solving for S reveals that $S = (r^{n+1} - 1)/(r - 1)$.

    3. Euclid derived this formula around 300 B.C.E.

C. The legend of the most "modest" mathematician.

    1. A king wished to reward a loyal mathematician and asked him what he wanted.

    2. The mathematician, who appeared both modest and humble, replied that if one grain of rice was placed on a square of an ordinary 8-by-8 chessboard and then two grains of rice were placed in the next square and so forth (doubling the previous amount of rice) until the last square was reached, then he would be content with the total sum of all the grains of rice.

    3. The king laughed and immediately granted this small request. However the king quickly stopped laughing—for the number of grains of rice owed to the mathematician equaled the following sum of terms from a geometric progression (recall that there are 64 squares on the chessboard): $1 + 2 + 2^2 + 2^3 + 2^4 + \cdots + 2^{63}$, which by our formula (with $r = 2$ and $n = 63$) equals: $(2^{64} - 1)/(2 - 1) = 18{,}446{,}744{,}073{,}709{,}551{,}615$ grains of rice.

    4. Given that a grain of rice weighs approximately 0.033 grams, this pile of rice would weigh approximately 671,023,802,629 tons.

    5. Needless to say, the king was faced with two choices—either give up his entire kingdom to the mathematician or have the mathematician executed. Guess who had the last laugh?

IV. Infinite geometric series and taxes.

    A. When does an infinite sum make sense?

        1. We recall that $1 + r + r^2 + r^3 + \cdots + r^n = (r^{n+1} - 1)/(r - 1)$.

        2. We now suppose that the ratio r is small, that is, $0 < r < 1$.

        3. As n gets larger and larger, r^n is getting smaller and smaller and is approaching 0.

        4. In this case we can consider the infinite sum of all the numbers in the geometric progression. This infinite sum is known as a *geometric series* and is extremely important in our study of number theory.

        5. A geometric series is an infinite sum of the form $1 + r + r^2 + r^3 + \cdots$. But does such an endless sum have a numerical value?

    B. Searching for a pattern.

        1. If we consider the geometric series $1 + 1/2 + (1/2)^2 + (1/2)^3 + (1/2)^4 + (1/2)^5 + \cdots$ as representing lengths of line segments, then we can see geometrically this infinite series equals 2. (Recall that we are assuming that $0 < r < 1$.)

        2. More generally, given that $1 + r + r^2 + r^3 + \cdots + r^n = (r^{n+1} - 1)/(r - 1)$, and r^{n+1} is approaching 0 as n gets larger and larger, we see that the infinite geometric series $1 + r + r^2 + r^3 + \cdots = 1/(1 - r)$.

        3. We can check this formula for the case $r = 1/2$ and see that the infinite series equals $1/(1 - 1/2) = 1/(1/2) = 2$, as we just saw.

        4. This formula for infinite geometric series will be an important and useful fact as we move into the subtle points of number theory.

    C. A prize-winning application.

        1. Suppose you win a million-dollar prize on a game show. At first you believe you have a million dollars to enjoy. However, Uncle Sam has other plans. He will take 1/3 of your bounty in tax.

        2. However, suppose the game show desires so much hype that it offers to pay the tax for you so you will take home the full \$1 million. So they pay $(1 + 1/3)$ million dollars. However, you still do not take home a million dollars, because you now have to pay 1/3 tax on the extra 1/3 they gave you (1/9 of a million dollars).

3. If they offer the extra 1/9 of a million dollars, then that additional amount will be taxed. How much must they offer so you can take home a million dollars after taxes?

4. The answer is the infinite series: $1 + (1/3) + (1/3)^2 + (1/3)^3 + (1/3)^4 + (1/3)^5 + \cdots = 1/(1 - 1/3) = 3/2$ million dollars. We see that the tax on this amount is $3/2 \times 1/3 = 1/2$ million dollars. If we deduct this from the 3/2 million, we see that you are left with exactly 1 million dollars—after taxes.

Questions to Consider:

1. Consider the geometric progression that begins −2, −10, −50, −250. Without explicitly adding the terms themselves, determine the sum of the first five terms. (*Hint*: Notice that this progression equals the progression you get if you start with 1, 5, 25, 125, … and then multiply each term by −2.)

2. Recall that the decimal number 0.999… means $9/10 + 9/100 + 9/1000 + \ldots$. Find the *sum* of this infinite geometric series using the formula introduced in the lecture. Are you surprised that your answer equals 0.999…? (*Hint*: Notice that this progression equals the progression you get if you start with $1 + 1/10 + 1/100 + \cdots$ and then multiply each term by 9/10.)

Lecture Four—Transcript
Geometric Progressions, Exponential Growth

We now enter the world of *elementary number theory*, which as I mentioned earlier, is not a euphemism for "easy number theory." Instead, it's an area of number theory that has its focus on fundamental questions about numbers, most of whose subtle answers do not involve advanced mathematical techniques. Here, with the notion fresh in our minds of an arithmetic progression—a list of numbers generated by successive addition—we open these lectures on elementary number theory by exploring the multiplicative cousin of arithmetic progressions.

These are known as *geometric progressions*—number lists generated by successive multiplication. A simple example of a geometric progression arises by the mere act of doubling. So 1, 2, 4, 8, 16, 32, and so forth, is an example of a geometric progression.

These numbers grow at a dramatically fast rate and possess enormous structure. We'll momentarily pause, later in this lecture, to consider how geometric progressions naturally appear in the world of music as a means of producing an even-tempered scale. This melodic excursion will involve notions that go back to the Pythagoreans and continue to confound piano tuners to this very day.

Next, we'll turn to an extremely important and useful object in more advanced number theory—the *sum* of the terms of a geometric progression. Our exploration into these sums of numbers will include some amusing ancient stories that hold important mathematical morals.

We'll then extend this addition of terms and consider the *endless* sum of *all* terms of certain geometric progressions, and make such vexing ideas intuitive by visualizing such sums geometrically. The endless sum of a geometric progression is known as a *geometric series*. These series are fundamental in all branches of mathematics and science, especially, as we'll see in future lectures, in number theory itself.

We'll then close this discussion with a whimsical illustration of the utility of infinite geometric series in the lives of lucky game-show contestants: the happy fantasy of winning a large cash prize, and the unpleasant reality of paying taxes on that newfound bounty.

Let's begin by recalling that an arithmetic progression is a list of numbers with the property that to get from any number to the next we need only add a fixed value. As we'll discover as our course unfolds, these arithmetic progressions, while simple in structure, play an important role in our number theory story.

Perhaps even more important, however, are the corresponding progressions in which the addition is replaced by multiplication. These progressions are called *geometric progressions*. To generate a geometric progression, we must be given the first number and then the constant multiple, which is known as the *ratio*, that generates the successive numbers on the list. For example, if we start with 1 and we're given a ratio of 2, then our geometric progression equals, well, we start with 1, and then we continually multiply our answer by 2 to get the next number. So 1; 2; multiplied by 2, we get 4; multiplied by 2, we get 8; multiplied by 2, we get 16; 32; 64; and so forth. It's easy to see that a general term in this geometric progression is given by 2 to the *n* power for some natural number *n*.

The constant multiple—in the previous example, it was 2—is called the *ratio* because the ratio of any two consecutive terms—the larger to the smaller—is always equal to that constant ratio. For example, notice that if we take a look at 32 and 16 that appear on our list and are adjacent, if I take 32 and divide it by 16, I get that constant ratio 2. Hence that constant is called the *ratio*.

In general, if we start with 1 and are given a ratio of, say, r, where r is just some number, then we could generate the geometric progression 1, r, r^2, r^3, r^4, r^5, and so on. We keep multiplying the previous answer by an extra factor of r.

These examples show us that a general term in a geometric progression that starts with 1 is of the form r^n for some whole number n. Since the varying quantity is the exponent n, if r is a number bigger than 1, we say that r^n *grows exponentially*. Very dramatic growth. If r is a small number, bigger than 0 but smaller than 1—so r is between 0 and 1—we say that r^n *decays exponentially*, because it shrinks down faster and faster.

For example, if we start with 1 and consider the ratio r equals 1/2, then we have a geometric progression that decays exponentially. We'd have 1, 1/2, 1/4, 1/8, 1/16, 1/32, and so forth. Notice those terms are getting smaller and smaller and smaller. That would be exponential decay in this case.

Let's take a moment to consider progressions in music and consider ratios of pitches. A musical interval is an *octave* if the two pitches have frequencies in a ratio of 2:1. An interval is a perfect fifth if the ratio of the two pitches is in a ratio of 3:2. So, for example, in modern Western music, the pitch A, above middle C, has a frequency of 440 Hertz, which is usually said "440 Hertz." Thus, the pitch A, one octave higher, has a frequency of 440×2, or 880 Hertz. The pitch E, one-fifth higher than A440, has a frequency of $440 \times 3/2$, which equals 660 Hertz.

In Western music, the *chromatic scale* begins at one pitch, say, A, and progresses up in what are called *half steps*, until it ends with the note that is one octave above the starting pitch. The pitches are derived from a progression of perfect fifths, starting with the first pitch.

A progression of perfect fifths is nothing more than a geometric progression with the ratio $r = 3/2$. So if we start at A440, we would have 440 and then $440 \times 3/2$, then $440 \times (3/2)^2$, then $440 \times (3/2)^3$, then $440 \times (3/2)^4$, and so on. In fact, the numbers we get would be 440, 660, 990, 1485, and so forth. Of course, these pitches are so high that only our pets would actually hear them, and so we really wouldn't be able to appreciate the music.

In order to keep the notes within the 440–880 Hertz range, we divide the frequencies by 2 in order to lower the pitches so that they fall into the correct octave. This process actually requires us to modify our attractive geometric progression, sadly. But happily, we could then actually hear the pitches that are generated.

Why do we end up with a 12-note chromatic scale? The answer to this question, which we'll discover for ourselves later in the course, involves irrational numbers, and we haven't even talked about those yet. So we'll leave this as a musical and mathematical cliffhanger for now, but we will return to the question, why 12?

Returning now to the geometric progressions that are the focus of this lecture, we note that it would be extremely useful to find the sum of the first few terms in a geometric progression, just as we did with arithmetic progressions.

If we consider the geometric progression 1, 3, 9, 27, 81, and so forth, here the common ratio is 3—we multiply by 3 each time. Then the first five successive sums would be 1, 4, 13, 40, 121. Where do these numbers come from? We just add up the terms in the geometric progression. For example, $1 + 3$ gives us the 4, $1 + 3 + 9$ gave us the 13, $1 + 3 + 9 + 27$ gives us the 40, and so forth.

A pattern is not immediately apparent, and you can see that we're moving into a more sophisticated world of number, because now the pattern doesn't just pop out. This is going to now require some thinking. So to build some intuition, we'll explore a particular example with some great care. Let's actually consider the example we just looked at where we have a constant multiple of 3—a ratio of 3. Let's let S be the sum of the first five terms, so S will equal $1 + 3 + 3^2 + 3^3 + 3^4$. So we have all those numbers, which, by the way, equals 121. Let's just keep that in the back of our heads.

Here's the wonderful trick that's going to allow us to simplify this object. Let's now multiply both sides of this equation by 3. When I multiply the left-hand side by 3, I'm left with $3S$. When I multiply the right-hand side by 3, we see that every single term in that sum gets hit with another factor of 3. If you notice what happens, the 1, when multiplied by a 3, equals 3. The 3, when multiplied by a 3, equals 3^2. It's kind of like those late-night programs where as the new guest comes on, the old guest shifts down. We just shift everyone down one. So I see $3 + 3^2 + 3^3 + 3^4$, and that last term, which originally was 3^4, when I multiply it by an extra factor of 3, becomes 3^5.

If we line these things up and now subtract the two equations, on the left-hand side, what do I have? I see $3S$, and I'm subtracting $1S$, so that's $(3 - 1) \times S$. On the right-hand side, notice that all the terms in the middle there—in the interior—align perfectly and actually add to give 0: $3 - 3$, $3^2 - 3^2$, $3^3 - 3^3$, $3^4 - 3^4$. All we're left with is on the very top, that 3^5, that dangling term way off to the right, and then on the bottom, we're subtracting that first 1. So I see $-1 + 3^5$.

We can find that $S = (3^5 - 1)/(3 - 1)$, which works out to be $(243 - 1)/2$, which happily is 121, as expected. We find something interesting here. What do we find?

What we find here is that for this particular sum, the answer is 3 raised to a power, where the power is one higher than the exponent appearing in the sum of terms, then we subtract off 1 and divide by $(3 - 1)$—the ratio minus 1. This is the pattern that we've been searching for, and we can check to see if this pattern holds for other sums.

For example, let's apply the analogous pattern with $1 + 3 + 3^2 + 3^3$. Well, we would conjecture that the sum would equal 3 to the—not the cube, but we actually increase the power by 1: 3^4. Then we subtract 1 and divide by $(3 - 1)$, which equals $(81 - 1)/2$, which is 40, which is exactly what we saw earlier.

We can now derive a formula for the more general sum: $1 + 3 + 3^2 + 3^3 + \cdots$ all the way out to 3^n power, where n is some unknown natural number. We'll again call this sum S, for convenience, and we'll just follow the exact same trick as before. One great thing about mathematicians is that once we have a great idea, we try to recycle it again and again.

So if we multiply this entire equation by 3, we can align most of those interior terms of S with most of the terms of $3S$, and then we could subtract, so we'd have $3S = 3 + 3^2 + 3^3 + \cdots$, out to 3^n plus that last term which is now 3^{n+1}, because we're multiplying that last original term by an extra factor of 3. That's the plus 1 in the exponent.

Underneath this, we just write down the S again. So S equals $1 + 3 + 3^2 \cdots$, and so forth, out to 3^n. Notice how everything aligns and the interior terms, when we subtract, I see $3S - S$, which is $(3 - 1)S$, and on the right-hand side, all the interior terms add to give 0, except for that very, very, very far right term, which is 3^{n+1}, and that very first term in the bottom sum, which I'm subtracting, so I see a -1. So I see that $(3 - 1)S = -1 + 3^{n+1}$.

Again, solving for S, we see that $S = (3^{n+1} - 1)/(3 - 1)$. We can now consider the corresponding sum for any ratio r.

Let's consider $1 + r + r^2$, and so forth, all the way down to r^n. Again, we'll call this sum S. If we multiply S by r, that's now the ratio—it used to be 3—we can align most of the terms in S with most of the terms in $r \times S$, just as before, and then subtract.

Look at the economy of ideas. We're just taking this one idea and looking at variations on the theme, the variations being the generalization. So we have $r \times S = r + r^2 + r^3 \cdots$, all the way out to $+ r^n$, plus the last term, r^{n+1}. Underneath it, aligned, we'll just write S without the extra multiple of r, so we see $S = 1 + r + r^2 + r^3 \cdots$, all the way out to $+ r^n$.

When we subtract, what do we have? On the left-hand side, we have r Ss and we're taking away 1 S, so that's $(r - 1)S$, and on the right, again, we see this enormous cancellation on the interior terms and we're left with $-1 + r^{n+1}$. Solving again for S, we see $S = (r^{n+1} - 1)/2(r - 1)$. So we see the general manifestation of the formulas we saw with the r equaling 3.

Euclid derived this formula around 300 B.C.E. So this formula goes back to antiquity.

I want to take a moment to share with you the legend of the most "modest" mathematician. Here we'll see another example of the sum of a geometric progression. This is a very old tale, which I really enjoy.

Quite a while ago, a king wished to reward a loyal mathematician and asked him what he desired. The mathematician, who by the way appeared both modest and humble, replied that if just 1 grain of rice was placed on a square of an ordinary 8-by-8 chessboard, and then 2 grains of rice in the next square, and so forth—doubling the previous amount of rice—until the last square on the chessboard was reached, then he would be totally content with the total sum of all the grains of rice.

The king laughed and immediately granted this seemingly small request. However, the king quickly stopped laughing, for the number of grains of rice owed to the mathematician equaled the sum of the terms from a geometric progression.

Let's take a look at this in action for a second. So, here we have a chessboard, and so the mathematician asked for 1 grain of rice in the first spot, and now we're considering the geometric progression where we double each time. So the next one would be 2 grains of rice, so there's the 2 grains of rice. If we double again, we'd see 4 grains of rice, and so here I have 4 grains of rice. The next square would consist of 8 grains of rice, and you can see that we have these grains of rice, and the next one would be 16, and so forth. We see the pattern.

Now, the mathematician wants the sum of all the grains of rice, so we're actually adding up a geometric progression. We're actually adding up $1 + 2 + 2^2 + 2^3 + 2^4$. Let's just think about it. What's the very last square, 2 to what power? We start with 1, and then we have 2 to the 1st power, then we have 2^2 in the third spot, we have 2^3 in the fourth spot, and so on. So what do we see? We see in the last spot, in the 64th spot, we're going to have 2 to 63rd power, because notice that we started with a 1 and then went to 2 to the 1st, 2 to the 2nd in the third spot, 2 to the 3rd power in the fourth spot, and so forth.

So in the 64th spot, we'll be at the power 63. We're adding $1 + 2 + 2^2 + 2^3 \cdots$, all the way out to $+ 2^{63}$. This is the geometric progression where we have $r = 2$ and we're going to sum all the terms up, up to $n = 63$. Our formula tells us the answer. It's 2 raised to the 64th power—that's $63 + 1$—then we subtract 1 and divide by $(2 - 1)$, which works out to be a 20-digit number. That's a lot of grains of rice.

How can we make sense of this extremely large amount of grains of rice? Given that a grain of rice weighs approximately 0.033 grams, this pile of rice would weigh more than 671 billion tons. To place this tonnage in perspective, today the world produces approximately a half a billion tons of rice each year. So this would take an awful long time just to grow, well over 1200 years.

Needless to say, the king was faced with two choices: either give up his entire kingdom to the mathematician to begin to pay off this enormous debt or simply have the mathematician executed. Guess who had the last laugh?

Anyway, it's an important moral, both about geometric progressions and how quickly they accumulate when you add them up, and also mathematicians shouldn't be so greedy, especially if they come across as being quite modest.

Finally, I want us to consider summing up infinitely many numbers from a geometric progression. First we must ask, when does an infinite sum of terms from a geometric progression even make sense? Let's begin by recalling that for any r that's not equal to 1, the formula for the sum of $1 + r + r^2 + r^3 \cdots$, all the way out to r^n, we derive to be $(r^{n+1} - 1)/(r - 1)$. That was the formula for the sum of the terms in the geometric progression that we found.

We now suppose that the ratio r is a small number, let's say a positive number that's less than 1. Thus let's think about it. As n gets larger and larger, r^n is going to be getting smaller and smaller, and is, in fact, approaching 0.

For example, if we let r equal 1/2, then let's look at the geometric progression. We have 1/2, 1/4, 1/8, 1/16, 1/32, 1/64, and you could see that as the denominator gets larger and larger, the actual term—the actual fraction—is getting smaller and smaller and smaller, and is heading to 0.

In this case, we can actually consider the infinite sum of all the numbers in this geometric progression. This infinite sum is known as a *geometric series* and is extremely important in our study of number theory, so we will come back to this again and again.

A geometric series is an infinite sum of the form $1 + r + r^2 + r^3 \cdots$, and so forth, but endlessly. Does such an endless sum have a numerical value? As always, we search for a pattern through an example.

Let's consider the geometric series $1 + 1/2 + (1/2)^2 + (1/2)^3 + (1/2)^4 + (1/2)^5 \cdots$, and so on, going on forever. If we view these individual terms as representing lengths of a line segment, each one representing a shorter and shorter line segment, then we can see geometrically that this infinite series equals 2, and I want to show that to you right now.

Here is a little piece of a number line, and you can think of this part here as being 0, this here as 1, and this point here as 2. If we look at the very first term in our sum, we actually see it's 1. So if we consider that length, a line segment of length 1—this silver pointer is going to represent our sum so far—we are at 1. So there's 1; it brings us to 1.

Now we add the next term in our series, which is 1/2. So now I have to go halfway to where I currently am, toward 0. Now I add 1/4, which is halfway. Now I add 1/8, which is halfway, and now I add 1/16, and now I add 1/32, and 1/64. I keep taking half and half and half, and if I do that forever, I will, in fact, fill up the entire line segment. So I see that that infinite series actually geometrically seems to fill up this line, so it has length 2. Therefore the series has a numerical value of 2.

Let's consider $1 + 1/2 + (1/2)^2 + (1/2)^3 \cdots$, and so forth, just up to 1/2 to the nth power. Well, we have the formula for that finite sum: It's $((1/2)^{n+1} - 1)/(1/2 - 1)$. Now recall that $(1/2)^{n+1}$ is approaching 0 as the n gets larger and larger and larger. Thus this formula is heading toward $-1/(1/2 - 1)$, which is $-1/(-1/2)$. The negative signs cancel each other out, and the reciprocal of 1/2 is actually 2, which agrees with our geometric analysis from a moment ago.

More generally, for any ratio r that's positive but less than 1, we recall that $1 + r + r^2 + r^3 \cdots$, and so forth, out to $+ r^n$, equals our formula, now familiar: $(r^{n+1} - 1)/(r - 1)$. The r^{n+1} is approaching 0 as n gets larger and larger, because remember, we're assuming that the r is a positive number that's less than 1. So higher powers will make the thing actually shrink.

Thus we see that the infinite series $1 + r + r^2 + r^3 \cdots$, and so forth, endlessly, will actually equal $1/(1 - r)$, just as we saw with the example of the ½'s.

We can apply this formula to find the value, say, of $1 + 1/5 + (1/5)^2 + (1/5)^3 \cdots$, and so forth. Here we see that the ratio $r = 1/5$, and so what does our formula tell us that this geometric series sums to? It would sum to $1/(1 - r)$, which is $1 - 1/5$. Well, $1 - 1/5$ is 4/5, so our sum is $1/(4/5)$, or the reciprocal of 4/5, which is 5/4. So that infinite series sums to a number, and that number is 5/4.

This formula for infinite geometric series is an important identity that we'll utilize as we move deeper and deeper into the world of number theory. So this identity for the sum of an infinite geometric progression—these geometric series—will come back again and again.

I want to close this lecture with a prize-winning application. Suppose that you win a million-dollar prize on a television game show. Congratulations, by the way. You first believe that you have a million dollars to enjoy, which sounds great. However, Uncle Sam has other plans for you. He will take 1/3 of your bounty in tax.

However, suppose now that the television network wants to generate so much hype that it offers to actually pay the tax for you. So you'll take home the full \$1 million. They say they'll give you $(1 + 1/3)$

million dollars, so that you could pay off the 1/3 to the government. However, you still don't take home a million dollars. Do you see why? Because you now have to pay 1/3 tax on the extra 1/3 million dollars they gave you. That's an extra 1/9 of a million dollars in tax that you have to pay additionally.

If they offered to pay that extra 1/9 of a million dollars, sounds good, but then that's an additional amount, which will be taxed as well. So how much must they offer you so that you actually, genuinely take home a million dollars after taxes? The answer is the infinite geometric series $1 + 1/3$—now they've got to pay for the tax on 1/3, which is $+ (1/3)^2$—now they have to pay a tax on the $(1/3)^2$, which is $+ (1/3)^3 + (1/3)^4 + (1/3)^5 \cdots$, and so on, forever, endlessly.

What does our formula tell us? Here we see the common ratio r is 1/3, so our series sums to $1/(1 - r)$, which in this case is 1/3. So we see $1/(1 - 1/3)$. That's $1/(2/3)$, and the reciprocal of 2/3 is 3/2. So 3/2 million dollars is what the television program has to provide. Actually, we can see that the tax on this amount is, well, we have 3/2 of a million dollars, we multiply that by a third, that gives us a half a million dollars in taxes we'd have to pay.

So if we deduct this amount from the 3/2 million that we have—3/2, by the way is 1½ million—we pay a 1/2 a million in tax, and *voilà*, we see we're left with exactly 1 million dollars after taxes.

The power of a geometric series is one that will be with us throughout our course. The fact that there's such an elegant formula that we ourselves can derive for that infinite sum of ever-shrinking terms in a geometric progression is certainly a work of beauty—and is certainly worth a million dollars.

Lecture Five
Recurrence Sequences

Scope:

As we have seen in the previous two lectures, we can generate both arithmetic and geometric progressions by either adding or multiplying a fixed value to a previous term in order to produce the next term in our progression. In this lecture we will extend these ideas by studying important patterns of numbers in which the next term in our number list is found by calculating a fixed, predetermined combination of the previous terms. We will see that both arithmetic and geometric progressions are very special examples of this much more general notion of number sequence. These more intricate number patterns are known as *recurrence sequences*. The most famous recurrence sequence is the list of Fibonacci numbers and their second cousins, the Lucas numbers. Using the Fibonacci and Lucas numbers as exemplars, we will explore the structure and patterns hidden within recurrence sequences. As we study their growth, we will come upon one of the most controversial and famous numbers in human history: the golden ratio. After discovering the arithmetic aesthetics of this number, we will apply its connection with Fibonacci numbers to reveal a very clever and practical method of converting between miles and kilometers.

Outline

I. Growing sequences with a starting seed and a simple rule.

    A. Extending the ideas of arithmetic and geometric progressions.

        1. For arithmetic progressions, we generate a new value by adding a given fixed number to the previous value in the progression.

        2. For geometric progressions, we generate a new value by multiplying a given fixed number by the previous value in the progression.

        3. We notice that both arithmetic and geometric progressions are generated by the initial number, known

as the *starting seed*, and the rule that produces the next number from the previous number.

B. The notion of recurrence.

    1. We now extend this method to produce more intricate lists of numbers.

    2. Instead of the simple rule of either adding or multiplying by a fixed number to generate the next term in our sequence of numbers, we will now consider more interesting combinations of the previous terms to generate the next number.

    3. Just as with arithmetic and geometric progressions, the new generating rule involving the previous numbers from our sequence will remain the same as we produce our list of numbers.

    4. Such sequences of numbers—number lists in which terms in the sequence are found by applying a fixed rule involving the numbers that came before—are called *recurrence sequences*.

C. The whole story from two pieces of information.

    1. As we have seen with both arithmetic and geometric progressions, a recurrence sequence can be described precisely by just giving the first few terms (the starting seeds) and then the fixed rule that generates the next number.

    2. Thus only two pieces of information are required to define a recurrence sequence: the starting seeds and the generating rule.

    3. So arithmetic and geometric progressions are each examples of recurrence sequences.

II. Patterns within the sums of Fibonacci and Lucas numbers.

  A. Two important illustrations.

    1. We now consider some concrete examples of recurrence sequences that are neither arithmetic nor geometric progressions.

    2. We fix our generating rule to be: Add two consecutive numbers to produce the next term in the sequence.

    3. We still require starting seeds. In this case, we need two numbers to allow our process to start.

4. Suppose that our starting seeds are 1 and 1. So the first two terms in our sequence are 1, 1, and the process to generate the next term is always the same—add the previous two numbers.

5. In this case our recurrence sequence is 1, 1, 2, 3, 5, 8, 13, 21, 34, 55, 89, 144, and so forth. This famous sequence is known as the *Fibonacci numbers*, named after the 13th-century Italian mathematician Leonardo de Pisa (also known as Fibonacci). These numbers appear in unexpected places, including nature and art. They will make cameo appearances throughout our course.

6. We can generate a different recurrence sequence by simply changing the starting seeds (but keeping the same rule of adding two adjacent numbers to generate the next number). Suppose we start with 2 and 1.

7. This new recurrence sequence begins 2, 1, 3, 4, 7, 11, 18, 29, 47, 76, 123, 199, and so forth. This sequence is called the *Lucas sequence*, named after the 19th-century French mathematician Edouard Lucas.

B. Hidden patterns within the sequences.
    1. As we march through these sequences of numbers, the two appear to have nothing in common. However, they do share the same generating rule, so perhaps there is some hidden connection.

    2. If we sum pairs of Fibonacci numbers that are two apart from each other we see: $1 + 2 = 3$; $1 + 3 = 4$; $2 + 5 = 7$; $3 + 8 = 11$. Incredibly, we discover the Lucas sequence.

    3. The idea of the inductive proof of this theorem can be seen by considering $3 + 8$ and first writing each as the sum of two adjacent Fibonacci numbers: $3 + 8 = (1 + 2) + (3 + 5) = (1 + 3) + (2 + 5) = 4 + 7$, which is the sum of two adjacent Lucas numbers. Therefore this sum equals the next Lucas number, 11.

    4. As before with the progressions we studied, we can consider the sum of the first terms of the Fibonacci or Lucas sequences. If we compute cumulative sums of the first terms of the Fibonacci sequence, we obtain: 1, 2, 4, 7, 12, 20, 33, and so forth. What do we see?

5. If we add 1 to each number on our list, we return to the Fibonacci sequence: 2, 3, 5, 8, 13, 21, 34, and so forth. Can this "magic" be explained by mathematics?

6. We can use a "domino idea": We consider a sum of Fibonacci numbers but add 1 at the beginning. That additional 1 generates a chain reaction of consecutive Fibonacci numbers:

$$1 + 1 + 1 + 2 + 3 + 5$$
$$= 2 + 1 + 2 + 3 + 5$$
$$= 3 + 2 + 3 + 5$$
$$= 5 + 3 + 5$$
$$= 8 + 5$$
$$= 13.$$

7. A similar result holds for the Lucas sequence as well.

III. Measuring the growth of recurrence sequences of numbers via ratios.

  A. Since the term we add to a Fibonacci number to generate the next one is increasing, we know that the Fibonacci sequence is *not* an arithmetic progression. But perhaps the Fibonacci sequence is a geometric progression.

  B. This issue inspires us to examine the *ratios* of consecutive Fibonacci numbers.

    1. $1/1$; $2/1 = (1 + 1)/1 = 1 + 1/1$; $3/2 = (2 + 1)/2 = 1 + 1/2 = 1 + 1/(1 + 1/1)$; $5/3 = (3 + 2)/3 = 1 + 2/3 = 1 + 1/(1 + 1/(1 + 1/1))$.

    2. Thus we see that while the Fibonacci sequence is not a geometric progression, the ratios of consecutive terms *are* approaching a number: An endless fraction-within-fraction of the form $1+ 1/(1 + 1/(1 + 1/(1+ 1/(1 + 1/(1 + 1/(1+ 1/(1 + 1/(1 + 1/(1 \cdots$.

    3. This number is called the *golden ratio*.

IV. The golden ratio.

  A. A continued fraction.

    1. This representation of the golden ratio as an endless fraction-within-fraction is an example of a *continued fraction*, which we will study at the end of this course. Thus this number foreshadows our number travels to come.

2. We now wonder if the golden ratio can be expressed as a more familiar-looking number.

B. A self-similar number.
1. We notice that if we call the golden ratio φ [phi] then we realize that $\varphi = 1 + 1/\varphi$.
2. Solving this equation is equivalent to solving the quadratic equation $\varphi^2 - \varphi - 1 = 0$.
3. Using the quadratic formula, we find two solutions: $(1 + \sqrt{5})/2$ and $(1 - \sqrt{5})/2$. Since the second solution is negative and φ is positive, we discover that $\varphi = (1 + \sqrt{5})/2$.
4. The decimal expansion for φ is 1.618033989..., which hides much of the structure of φ that we already uncovered.

C. Aesthetics and beauty within number.
1. While φ is one of the most famous numbers, it is also one of the most controversial. Many believe that the golden ratio appears in nature and art and informs our aesthetic tastes. Others disagree.
2. It is a fact that the golden ratio does appear in many different areas of number theory, mathematics, and science.
3. It is also a fact that the golden ratio is the only number that can be expressed as an endless continued fraction consisting solely of 1s.
4. We will return to the golden ratio and see that even though it is an irrational number, it is the "least" irrational number that exists. Thus within the world of number theory, it is difficult to argue that φ is not an abstract object of great importance and beauty.

D. Using φ today in our everyday travels.
1. We close this lecture with a practical application of the Fibonacci numbers and their connection with the golden ratio.
2. The golden ratio equals 1.618..., which coincidentally is extremely close to the number of kilometers that equals a mile. In fact, 1 mile = 1.6093... kilometers.
3. We can use the Fibonacci numbers to convert between kilometers and miles.

4. To illustrate the method, we will approximate the number of miles in a 10-km run. First we express 10 as a sum of Fibonacci numbers: $10 = 2 + 8$. Next we replace each Fibonacci number in our sum with the Fibonacci number that precedes it in the Fibonacci sequence. In this case we replace $2 + 8$ by $1 + 5$, which equals 6. Therefore 10 km is approximately 6 miles (in fact, it is about 6.2 miles). How many miles is a 50-km trip? We write $50 = 34 + 13 + 3$ and then compute $21 + 8 + 2 = 31$ miles (actual mileage is 31.06…).

Questions to Consider:

1. a) Without adding them directly, determine the sum of the first 10 Fibonacci numbers.

b) Use the Fibonacci sequence to find the number of kilometers roughly equivalent to 100 miles. (*Hint*: First write 100 as the sum of Fibonacci numbers.)

2. Look at attractive rectangular shapes around you. For each one, compute the ratio of the longer side to the shorter side. Are any of your ratios close to the golden ratio?

Lecture Five—Transcript
Recurrence Sequences

As we've seen in the previous two lectures, we can generate both arithmetic and geometric progressions by either adding or multiplying by a fixed value to a previous term in order to produce the next term in our progression.

In this lecture, we'll extend these ideas by studying important patterns of numbers in which the next term in our number list is found by calculating a fixed, predetermined combination of the previous terms. We'll see that both arithmetic and geometric progressions are very special examples of this much more general notion of number sequence.

These more intricate number patterns are known as *recurrence sequences*. The most famous recurrence sequence is the list of Fibonacci numbers and their second cousins, the Lucas numbers. If you're a Fibonacci fan, then you'll be delighted to know that the Fibonacci numbers will make several surprise cameo appearances throughout our course. If you're not currently a Fibonacci fan, then I assure you that you will be one by the end of our number theory journey.

Using the Fibonacci and Lucas numbers as exemplars, we'll explore the structure and patterns hidden within recurrence sequences. As we study their growth, we'll come upon one of the most controversial and famous numbers in human history: the golden ratio. After discovering the arithmetic aesthetics of this number, we'll apply its connection with Fibonacci numbers to reveal a very clever and practical method of converting between miles and kilometers.

Here we open by extending the ideas of arithmetic and geometric progressions. For arithmetic progressions, we generate a new value by adding a given fixed number to the previous value in the progression. For a geometric progression, we generate a new value by multiplying a given fixed number by the previous value in our progression.

Let's notice that both arithmetic and geometric progressions share a common trait. They are generated by the initial number, which is known as the *starting seed* because everything grows from there, and then the rule that produces the next number from the previous number.

We now extend this trait to produce more intricate lists of numbers. Instead of the simple rule of either adding or multiplying by a fixed number to generate the next number in our sequence of numbers, we'll now consider more interesting combinations of the previous terms to generate the next number.

Just as with arithmetic and geometric progressions, the new generating rule involving the previous numbers from our sequence will remain the same as we generate the terms in our sequence of numbers. So we have this fixed rule that will follow us throughout our sequence.

The number lists in which the terms are found by applying a fixed rule involving the numbers that came before are called *recurrence sequences*. As we've seen with both arithmetic and geometric progressions, a recurrence sequence can be described precisely by just giving the first few terms—the starting seeds—and then the fixed rule that generates the next term. That's all that's required: the first starting seeds and then the rule. Everything else then is determined.

Therefore, only two pieces of information are required to define a recurrence sequence: the starting seed and the generating rule. In fact, both arithmetic and geometric progressions are each examples of recurrence sequences.

In order to ground our discussion, let's consider two important illustrations. So, now we're going to get our feet to the ground. The concrete examples of recurrence sequences that we're about to consider are neither arithmetic nor geometric progressions. We're going to generate something that's genuinely new for us in terms of our work together so far in the course.

Let's fix our generating rule to be that we add two consecutive numbers on our list in order to produce the next term in our sequence. Now we're going to actually require starting seeds, but we're going to require two starting seeds, because the rule says that we take the two elements that are here, adjacent, and add them to get the next one. Then we add these two elements to get the next one, add these two elements to get the next one, and so forth. So we're still requiring starting seeds, but in this case we're going to actually need two in order to allow our process to genuinely get off the ground.

Suppose that our starting seeds were simply 1, followed by a 1. So those are the first two elements in our progression. Now we're going to apply this rule. So the process to generate the next term is to always do the same thing, which is to add the two previous numbers. In our case, the recurrence sequence would be, well, 1, followed by 1, those are our starting numbers, and then we add those two together to get 2, that's the next one. Then we add the 1 and the 2, add them together and get a 3. We now add the 2 and the 3, and add them together to get a 5. If we continue, we see 8, followed by 13, 21, 34, 55, 89, 144, and so on.

This famous sequence is known as the *Fibonacci numbers*, named after the 13th-century Italian mathematician Leonardo de Pisa, who was also known as Fibonacci. These numbers appear in unexpected places, including nature and art, and they'll make several appearances throughout our course as well.

We can generate a different recurrence sequence by just simply changing the starting seeds, but keeping the same pattern—the same rule—of adding two adjacent numbers to generate the next number. Now, let's suppose we start with 2 and 1 as our starting seeds. Then the new recurrence sequence begins 2, 1, and then we add those two together to get 3, and then we add the 1 and the 3, and add them together to get the 4, the 3 and the 4 get added to get the 7, and the next one would be 11, then 18, 29, 47, 76, 123, 199, and so on. The list goes on. This sequence is called the *Lucas sequence*, named after the 19th-century French mathematician Edouard Lucas.

As we march through these sequences of numbers, especially as we walk through them as they get larger and larger, the two lists appear to have nothing in common. However, they do share the same generating rule: adding the two previous numbers to get the next one. Perhaps there's some hidden connection, some structure that genuinely connects these two sequences together. Let's see what we can find.

Remember, the thing that we do in number theory is to try to explore and then uncover a hidden pattern, which leads to hidden structure.

We know that if we sum two consecutive Fibonacci numbers, we get the next one on the list. What if we sum pairs of Fibonacci numbers that are two apart from each other? That is to say, instead of adding these two to get the next one, let's add this one and this one—skipping this one—and just see what we get. For example, if we take 1 and skip the second 1 and go right to the 2, if we add those, $1 + 2$, we get 3. If we add the 1 and the 3, we get 4. If we take a look at the Fibonacci number 2, skip the 3 and go to the next one, which is 5, then $2 + 5$ is 7. The next pair would be $3 + 8$, which is 11, so what do we see? We see 3, 4, 7, 11—and what do we see? Incredibly, we're generating the Lucas sequence. Absolutely unbelievable.

Let's see why this amazing phenomenon actually occurs. I want us to look at an example very closely so we can actually see what's hidden behind the scenes. Let's look at the example $3 + 8$. Of course, we could just add it up and say, "3 plus 8 is 11," and be done with it, but I'm trying to find a pattern that's going to work even when the numbers aren't 3 and 8.

The first thing we're going to do is use the very definition of Fibonacci numbers. So what we're going to do is we're going to write 3 as the sum of the two Fibonacci numbers that come right before it, so the two adjacent Fibonacci numbers. Then we'll take 8 and write it by its defining trait, namely, the sum of the two previous Fibonacci numbers.

In this case, we'd see that $3 + 8$ equals $(1 + 2)$—that's the 3 part—+ $(3 + 5)$—that's the 8 part. Notice I'm using the generating rule for the Fibonacci numbers in order to rewrite the 3 and rewrite the 8 into smaller numbers. We're going to use a "divide and conquer" approach; we're going to try to make the numbers smaller and analyze them.

Instead of writing $3 + 8$, we're going to have $(1 + 2) + (3 + 5)$. Now I'm going to perform a very clever trick. I'm going to take those two numbers on the inside of the sum, the 2 and the 3, and I'm going to switch the order of them. Now I'm going to write $(1 + 3) + (2 + 5)$. Of course, with addition, it doesn't matter what order we add, so this is still equal to the original $3 + 8$.

Notice what happens when I group the first two numbers together and now the last two numbers together and add them. I no longer get the 3 plus the 8; I get something else. I now get $4 + 7$. Notice that both 4 and 7 are Lucas numbers. In fact, more is true: They're actually adjacent Lucas numbers.

So, what do we know? If we have two adjacent Lucas numbers and we add them together, their sum is going to equal another Lucas number, in this case, 11.

That was the key idea—to write the two numbers as their individual sums and then switch these, and all of a sudden we see two adjacent Lucas numbers. When we add them up, we get a Lucas number. When we add a pair of Fibonacci numbers that are two apart in the sequence, we see that the sum is indeed a Lucas number.

This example can be extended to show that the result holds in general, and a rigorous proof involves a technique known as *mathematical induction*, which is basically, in some sense, this "divide and conquer" approach. The method involves using the smaller cases to show the validity of the result for the larger cases, just as our example really illustrated.

As in the previous lectures with the progressions that we studied earlier, we could consider the sum of the first few terms of the Fibonacci or Lucas sequences. So let's consider them right now. If we compute the cumulative sums of the first terms of the Fibonacci sequence, what would we see? Well, we'd see 1, that's just the first term, then $1 + 1$, $1 + 1 + 2$, $1 + 1 + 2 + 3$, $1 + 1 + 2 + 3 + 5$, and so on. What are those sums? Those sums equal 1, 2, 4, 7, 12, 20, 33, and so on.

What do we see when we look at these numbers? The answer is not much. That's okay. It means that maybe something more subtle is going on. What if we take this list of numbers that we just found, this sum, and what if we add 1, the number 1, to each number on our list? So, I'm going to increase every number on this sum list by 1. Then what do we see?

We see 2, 3, 5, 8, 13, 21, 34, and like magic we return to the Fibonacci sequence. Absolutely amazing. Can this "magic" be explained by mathematics? The answer is yes, and we actually are going to use what I think of as a "domino idea." We're going to consider the sum of all the Fibonacci numbers up to a certain point, and then what I'm going to do is I'm going to add that extra 1 that we're supposed to add, at the very, very beginning—at the very, very beginning. Let's see what happens.

What's going to happen, by the way, is that that 1 at the very beginning is going to create a chain reaction, and all of a sudden the sums are going to implode into Fibonacci numbers. And we're going to get this entire collapsing down to just one Fibonacci number, as we've seen in the illustration of the examples we've looked at.

So let's actually consider an example. Let's take $1 + 1 + 2 + 3 + 5$—that's the sum of the first few Fibonacci numbers—and then we're going to stick on a 1 at the very, very front of that. Now what do we see? Let's just add the first two numbers together. That's the new 1 plus the first 1 that we had in the Fibonacci list. What do we see? One plus 1 is 2. Notice that 2 is, in fact, a Fibonacci number.

What's the next number in our sum now? Well, it's a 1. Notice that 2 and 1 are adjacent Fibonacci numbers, so their sum—which is 3—yields another Fibonacci number. So, now we have $3 + 2 + 3 + 5$. Well, $3 + 2$, those are two adjacent Fibonacci numbers, so when we add them together, we get the next Fibonacci number, which in this case is 5. So now our sum has collapsed down to $5 + 3 + 5$, and notice that $5 + 3$, well, that generates the next Fibonacci number, because 5 and 3 are adjacent Fibonacci numbers. So I see 8. So, I have $8 + 5$; $8 + 5$, we have, again, adjacent Fibonacci numbers, so the sum is 13.

This is the idea, and a similar result actually holds for the Lucas sequence as well, which you're free to think about if you want. But the point here is that when we collapse this down, we see two adjacent Fibonacci numbers; however, in this instance, the adjacent Fibonacci numbers are kind of reversed in order. It's first bigger plus smaller, but still adjacent so the sum is a Fibonacci number.

We now turn to the growth of recurrence sequences of numbers. How are these numbers actually growing? Since the term we add to one Fibonacci number to generate the next one is ever increasing—we're not adding a fixed amount, we're always adding a number that comes right before the previous Fibonacci number in order to generate the next one—we know that the Fibonacci sequence is *not* an arithmetic progression. But perhaps the Fibonacci sequence is a geometric progression. That's possible.

Remember that a geometric progression is one in which the ratio of consecutive terms in the list is always the same. So, let's look at the first few ratios and see if it's constant: 1 divided by 1—the ratio of the first two terms—equals 1. The ratio of the next two terms: 2 divided by 1 is 2. That's not 1, so we're already done. Let's just take one more example: 3 divided by 2 is 1.5. Clearly these ratios are not equal, so we see that the Fibonacci numbers do not form a geometric progression.

But these numbers are near each other. We got 1, 2, 1.5, they're all kind of hanging around the same area, so maybe there's something else going on within these ratios that requires a closer look.

This time, we're going to write the ratios in a different way, one that actually captures the generating rule for producing the Fibonacci numbers. Let's look at the ratios again. We have 1/1, which I'm just going to keep as 1/1. I'm not going to simplify it to just the number 1.

The next ratio is 2/1, and how did we get the 2? We got the 2 by adding 1 + 1. So let's now write the 2 as (1 + 1)/1, which, notice the first 1/1 is just the number 1 (there's a lot of 1s here, by the way), + 1/1. So 2 actually can be written as (1 + 1)/1. Probably the most complicated way of writing 2 you'll ever see, ever. But I want us to write it this way because I'm searching for a pattern.

Let's consider the next ratio: 3/2. Again, I'm going to apply my rule of writing how 3 was born: 3 was born by taking 2 and adding 1 to it, so we see that 3 = 2 + 1. We're dividing that all by 2, so we see 2/2 + 1/2. Well, 2/2 is just the number 1, so we have 1 + 1/2. Notice that 1/2 is the reciprocal of 2, which was our previous ratio. So if I take the reciprocal of the previous ratio, that's equal to 1/2, and so I see that 3/2 is 1 + 1/(1 + 1/1). Interesting. Lots of (1 + 1/1)s.

Let's look at 5/3 now. Well 5/3, I'm going to just follow the same process. I'm going to write 5 by its defining trait within the Fibonacci list as 3 + 2. Then when I break up 3 + 2, all divided by 3, into two fractions, I see 3/3 + 2/3. Well 3/3 is just 1, so I see 1 + 2/3. But notice again that 2/3 is the reciprocal of the previous answer we found: 3/2. Therefore, I could take the previous answer and just take its reciprocal to get that second term.

So I see that 5/3 equals 1 + 1/(1 + 1/(1 + 1/1)). Amazing.

Even though these ratios are not constant, they do have a predictable and even fascinating form. They're an ever-deepening expansion of fractions of the form 1 + 1/(1 + (1/(1 + (1/…, and so forth.

If we now imagine continuing this pattern endlessly, then we arrive at an endless fraction-within-fraction of the form $1 + 1/(1 + (1/(1 + (1/(1 + (1/...$, forever.

As improbable as it may seem, this expansion has a numerical value, and it's actually a very special number. This number is called the *golden ratio*. This representation of the golden ratio as an endless fraction-within-fraction is actually an example of what we call a *continued fraction*, which we'll study in some depth at the very end of this course.

We now wonder if the golden ratio can be expressed as a more familiar-looking number. We notice that if we call the golden ratio by its normal name, which we use the Greek letter φ [phi], then we have that φ equals—and now I want us to visualize this: φ equals $1 + 1/(1 + (1/(1 + (1/(1 + (1/...$, endlessly. That's the definition of φ. Whenever we see an endless list of $1 + 1/(1 + (1/(1 + (1/...$, endlessly, we know that that actually equals φ.

Now I'm going to perform a little magic trick. Let's take that expression and just cover up the first $1 + 1/$. What's left right here? Well, it's actually an endless list of $1 + 1/(1 + (1/(1 + (1/...$, and so forth, forever. So, that piece right here is actually another copy of the golden ratio. It's another copy of φ. What do I see? I see that φ equals $1 + 1/\varphi$. So φ equals $1 + 1$ over itself. It's almost a peculiar identity that we're finding because we're seeing that φ actually is inside of itself in some crazy way.

The important thing for us is that this actually can be viewed as an equation, and now we can solve for this number, φ. If we do that, I'll first multiply both sides by φ because no one likes having an unknown in the denominator. So I'd see $\varphi^2 = \varphi + 1$, because φ/φ would just be 1.

Now I see φ^2. That means it's a quadratic, which if you remember from your old school days, the way to solve a quadratic is to have everybody go to one side. It's a big party at my house, come on over, and we have to have a thing equal to 0. So then I'd see $\varphi^2 - \varphi - 1$—if I subtract the left-hand side over—then we'd see simply $\varphi^2 - \varphi - 1 = 0$.

Using the quadratic formula, which we saw in grade school, we can actually find the two solutions to this. There will be two solutions because it's a quadratic. That exponent is a 2, which means we'll have two answers.

One solution is $(1 + \sqrt{5})/2$, and the other solution is $(1 - \sqrt{5})/2$. That second number, $(1 - \sqrt{5})/2$, is actually a negative number, and since our number φ starts off with 1 +, and some positive number, we know it's going to be a positive answer. Therefore we use the other solution, and we just discovered that φ—the golden ratio—equals $(1 + \sqrt{5})/2$.

The decimal expansion for φ, if you now plug this into a calculator, $(1 + \sqrt{5})/2$, we'd see that φ equals 1.618033989…, and it goes on, and so forth. Which, by the way, when viewed as a decimal number, just seems like, almost like a jumbled, random screed of digits, and we completely see that hidden from sight is the structure of φ that we've already uncovered. So a great life lesson, again, which we've seen time and time again is that when we look at things in a different light, all of a sudden we might see structure, whereas another view might hide it.

We now see that the ratio of any consecutive Fibonacci numbers is approximately equal to this number, φ: 1.618033…, and so forth. The larger the Fibonacci numbers get, the closer the approximation is, in fact, to this golden ratio.

While φ is one of the most famous numbers, it is also one of the most controversial. Many believe that the golden ratio informs our aesthetic tastes, appearing in nature and art, while others strongly disagree. In fact, this is quite a big bone of contention amongst many people: fans of numerology, fans of number theory, and fans of the arts.

For example, a famous example that people sometimes point to is the Parthenon and the Acropolis, and whether within the Parthenon, if we look at the dimensions and look at the rectangle formed by circumscribing a rectangle around the Parthenon, if you look at the ratio of the base length to the height, that that ratio would be the golden ratio.

Rectangles that have this property, a rectangle that has the property that a ratio between the length of the base to the length of the height is exactly the golden ratio, are referred to as golden rectangles. So the question is, do we see a golden rectangle within the Parthenon? The answer is probably no, because of course this number is infinitely complicated. The decimal expansion goes on forever, so it's hard to measure exactly. But are we drawn to that particular type of shape? Are we drawn to the aesthetics of this number? Many people believe no, many people believe yes, and so it remains a wonderful debate.

In fact, the golden ratio does appear in many different areas of number theory and science, and that remains a fact. The number is important in our work in mathematics and science. It's also a fact that the golden ratio is the only number that can be expressed as such an endless continued fraction, consisting solely of 1s, in the fashion $1 + (1/(1 + (1/(1 + (1/\ldots$, forever.

By the way, we'll return to the golden ratio later in this course and see that it actually has a very important role to play in our study of irrational numbers. Certainly within the world of number theory, it would be difficult to argue that φ—the golden ratio—is not an abstract object of great importance and beauty.

While we can debate whether it appears in nature and in aesthetics, it certainly appears in mathematics, and it certainly, within that realm, is an aesthetically appealing and important number.

I want to close this lecture with a practical application of the Fibonacci numbers and their connection with the golden ratio, so here's an actual thing we can use in our everyday lives.

Recall that we found the value of the golden ratio to be around 1.618, which coincidentally is extremely close to the number of kilometers that equals a mile. In fact, 1 mile equals 1.6093-something kilometers, which is really close to 1.61-something.

We can actually use the Fibonacci numbers to convert between kilometers and miles. Let me illustrate this with an example. For example, 13 kilometers is approximately 8 miles, since 13/8 is approximately 1.6. How did I get that? Since 13 is a Fibonacci number, all I did was find the Fibonacci number that comes right before it, which is 8. We know that that ratio, 13/8, is going to be very close to the golden ratio, which is very close to the conversion ratio for kilometers to miles. What I see here is that 13 kilometers is approximately 8 miles. If you have a Fibonacci number of kilometers, to convert it to miles just look at the Fibonacci number that comes right before.

To illustrate the method in general, let's now try to approximate the number of miles in a 10-kilometer run. First, we express 10 as a sum of Fibonacci numbers. In this case, notice that 10 equals 2 + 8. Next, we just follow the procedure we just saw. We replace each Fibonacci number in our sum with the Fibonacci number that precedes it in the Fibonacci sequence.

In this case, we replace the 2 by a 1, and the 8 by a 5, and so instead of 2 + 8 we now consider 1 + 5, and if we add 1 and 5 we get 6. Therefore, 10 kilometers is approximately 6 miles. If you've ever run in a 10-kilometer race, you know that's right, because in fact a 10-kilometer race is about 6.2 miles. So really, we have a close estimate, just using the Fibonacci numbers.

How long is a 50-kilometer trip in miles? We first write 50 as the sum of Fibonacci numbers, which by the way you can always do by taking a look at 50 and backing up until you see the very first Fibonacci number that you come to, which in this case would be 34, and then you'd be left with 16, and so you keep backing up until you hit the next Fibonacci number, which would be 13, and then you back up more until you get to 3.

So, 50 = 34 + 13 + 3. Now if we take each Fibonacci number and replace it by the Fibonacci number that comes right before, we would compute 21 (replacing the 34), the 13 would get replaced by 8, and the 3 would get replaced by a 2, and so we now see our sum would be 21 + 8 + 2, which is 31 miles.

In actuality, the mileage conversion of a 50-kilometer trip is 31.06 miles. So it's incredibly close to 31. You could see how wonderful this approximation really is.

Practicality aside, we've now seen the world of recurrence sequences, which takes the foundational ideas we saw with arithmetic progressions and geometric progressions and extends it to a wide range of important number lists that include the Fibonacci and Lucas numbers.

Lecture Six
The Binet Formula and the Towers of Hanoi

Scope:

Beyond their intrinsic appeal and utility to number theorists, recurrence sequences of numbers are important objects of computer science. The question of considerable interest is, Can we find a formula that will produce any individual term in this sequence of numbers without the need for generating *all* the numbers in the list up to that term? In this lecture we will tackle this challenge by discovering the famous Binet formula for the Fibonacci numbers. While named after the French mathematician Jacques Binet who first derived it in 1843, it appears that this important formula was apparently known to Leonhard Euler and Daniel Bernoulli over 100 years earlier. Once we derive this formula, by separating a pattern we will realize that our method can be generalized and used to find corresponding formulas for all such recurrence sequences. The Binet formula will provide us with the insight that while recurrence sequences such as the Fibonacci and Lucas numbers are not geometric progressions, they are in fact a combination of two geometric progressions. We will then close this lecture with one of the most famous stories involving recurrence sequences of numbers: The Towers of Hanoi.

Outline

I. The practical importance of recurrence sequences.

    A. A world of recurrence.

        1. In the previous lecture we discovered recurrence sequences: lists of numbers that can be generated using some starting seeds (the first few numbers) and a rule for generating future numbers.

        2. Arithmetic and geometric progressions are very simple examples of recurrence sequences, as are the Fibonacci and Lucas sequences.

    B. An important idea within computer science.

        1. A recurrence relation is one in which previous information is used in a systematic manner to generate new information.

2. The concept of recurrence is an important and fundamental component in many computer algorithms and languages.
3. As a result, all computer scientists have studied and use recurrence sequences in their programming.

II. Finding a "closed formula" for the Fibonacci numbers.

A. A recurrence definition versus a formula.

1. Suppose we let F_n denote the nth Fibonacci number. So we have $F_1 = 1$, $F_2 = 1$, $F_3 = 2$, $F_4 = 3$, $F_5 = 5$, $F_6 = 8$, $F_7 = 13$, and so forth.
2. This notation allows us to precisely define the recurrence rule as $F_{n+1} = F_n + F_{n-1}$, for all $n \geq 2$.
3. One disadvantage to describing the Fibonacci numbers in this manner is that if we wish to find the value of the 1000th Fibonacci number we would be required to compute all the previous ones in succession and work our way to the 1000th one.
4. Ideally we would like a "closed formula"—that is, a generic formula in which we can just plug in 1000 to produce the 1000th Fibonacci number without computing any others.
5. One of the important features of many recurrence sequences, in general, is that such closed formulas can be derived, and we will illustrate this process with the Fibonacci numbers.

B. A focus on φ.

1. We recall that in our attempt to express φ in a more familiar form (rather than as a continued fraction), we saw that φ was one of the solutions to $x^2 = x + 1$.
2. Using the quadratic formula we found that the two solutions to this equation are $(1 + \sqrt{5})/2$ and $(1 - \sqrt{5})/2$. Because φ is a positive number, we found that $\varphi = (1 + \sqrt{5})/2$. We now write τ for the negative solution, that is, $\tau = (1 - \sqrt{5})/2$.
3. Since both φ and τ are solutions to the equation, we see that $\varphi^2 = \varphi + 1$ and $\tau^2 = \tau + 1$.

4. We can use the formula $\varphi^2 = \varphi + 1$ to simplify φ^3: $\varphi^3 = \varphi\varphi^2 = \varphi(\varphi + 1) = \varphi^2 + \varphi = (\varphi + 1) + \varphi = 2\varphi + 1$. So we see that $\varphi^3 = 2\varphi + 1$.

C. Discovering a pattern.

    1. We apply the same technique to find φ^4: $\varphi^4 = \varphi\varphi^3 = \varphi(2\varphi + 1) = 2\varphi^2 + \varphi = 2(\varphi + 1) + \varphi = 3\varphi + 2$.

    2. Similarly for φ^5: $\varphi^5 = \varphi\varphi^4 = \varphi(3\varphi + 2) = 3\varphi^2 + 2\varphi = 3(\varphi + 1) + 2\varphi = 5\varphi + 3$.

    3. Finally we find $\varphi^6 = \varphi\varphi^5 = \varphi(5\varphi + 3) = 5\varphi^2 + 3\varphi = 5(\varphi + 1) + 3\varphi = 8\varphi + 5$, and a pattern emerges.

    4. We summarize our findings:

$$\varphi^2 = \varphi + 1$$
$$\varphi^3 = 2\varphi + 1$$
$$\varphi^4 = 3\varphi + 2$$
$$\varphi^5 = 5\varphi + 3$$
$$\varphi^6 = 8\varphi + 5.$$

    5. We see the Fibonacci numbers appearing. In fact we can continue this process indefinitely and thus conclude that in general for any natural number n, $\varphi^n = F_n\varphi + F_{n-1}$.

    6. By the identical reasoning, since τ is also a solution to $x^2 = x + 1$, we see that a corresponding amazing formula holds for τ. That is, for any natural number n, $\tau^n = F_n\tau + F_{n-1}$.

    7. If we now subtract these two formulas, we see:

$$\varphi^n = F_n\varphi + F_{n-1}$$
$$-\ \tau^n = F_n\tau + F_{n-1}$$
$$\overline{(\varphi^n - \tau^n) = F_n(\varphi - \tau) + 0.}$$

    8. We can now solve this equation for F_n and find that for all n, $F_n = (\varphi^n - \tau^n)/(\varphi - \tau)$.

    9. We note that $\varphi - \tau = (1 + \sqrt{5})/2 - (1 - \sqrt{5})/2 = \sqrt{5}$, and thus we have derived a closed formula for the nth Fibonacci number: $F_n = (\varphi^n - \tau^n)/\sqrt{5}$. This elegant formula today is known as the *Binet formula*, named after the 19th-century French mathematician Jacques Binet.

10. The 1000th Fibonacci number equals $(\varphi^{1000} - \tau^{1000})/\sqrt{5}$, which a computer can simplify to the 209-digit Fibonacci number:

> 43466557686937456435688527675040625802564660517371780402481729089536555417949051890403879840079255169295922593080322634775209689623239873322471161642996440906533187938298969649928516003704476137795166849228875.

D. Nearly a geometric progression.

    1. Recall that we found that the Fibonacci sequence is not a geometric progression since the ratio of consecutive terms is not constant.

    2. We did see that those ratios are converging on a particular value. The value the ratios are approaching is the golden ratio, φ.

    3. Using Binet's formula, we now discover that the Fibonacci numbers are, in fact, the difference of two geometric progressions.

E. Lucas numbers revealed.

    1. Applying our previous analysis with the Lucas sequence, we can derive the closed formula: For all $n > 1$, $L_n = \varphi^{n-1} + \tau^{n-1}$.

    2. In fact, the terms in any recurrence sequence can be expressed as a closed formula.

III. The legend of the Towers of Hanoi.

    A. The history of a towering tale.

        1. "The Towers of Hanoi" was a logic puzzle that was marketed in 1883 by a "Professor Claus." However, "Professor Claus" is, in fact, an anagram of its true inventor, Professor Lucas.

        2. The Towers of Hanoi puzzle consists of three pegs and a collection of punctured disks of different diameters that can be placed on the pegs.

        3. The puzzle begins with all the disks on one peg in order of diameter, with the largest disk on the bottom.

        4. The object is to transfer all the disks to another peg so that they end up residing on this new peg in the original descending order. The rules are: Only one disk can be moved from one peg to another at a time, and at no time can a larger disk be placed on top of a smaller disk.

    5. Our challenge is to find a method for moving the disks and to determine the number of moves required.

B. A towering recurrence.

    1. Suppose we have n disks to be moved. We begin by observing that to move the largest (very bottom) disk, we must first move the other $n - 1$ smaller disks.

    2. Once we move those $n - 1$ disks to a new peg, then we can remove the last, largest disk and move it to the remaining empty peg. Now we must move the other $n - 1$ smaller disks back on top of the largest one.

    3. Let us write h_n for the number of moves required to move n disks. We now wonder if we can find a recurrence rule that generates the sequence of h_n's.

C. Discovering a pattern.

    1. By experimenting we can see that $h_1 = 1$, $h_2 = 3$, and $h_3 = 7$.

    2. In view of the disk-moving process, h_n must equal $h_{n-1} + h_{n-1} + 1$, that is, we have the recurrence rule: $h_n = 2h_{n-1} + 1$.

    3. So the number of moves required for 4 disks equals $2 \times 7 + 1$, which equals 15.

    4. We can find a closed formula by looking at our sequence of h_n's: 1, 3, 7, 15. We notice that each number is one less than a power of two: $1 = 2 - 1$, $3 = 2^2 - 1$, $7 = 2^3 - 1$, and $15 = 2^4 - 1$.

    5. In general, it is possible to prove that $h_n = 2^n - 1$, which gives a simple closed formula for the number of disk moves required if we have n disks.

D. Determining the end of the world.

    1. There is a legend that a certain group of monks had a particularly impressive edition of this puzzle consisting of 64 gold disks and three diamond pegs. They were able to move 1 disk per second.

    2. The legend is that the world would end once the monks completed their mission. We can now employ the closed formula we just found to predict when the world will end.

    3. The number of moves required to solve this puzzle with 64 disks equals $2^{64} - 1$. Given that the monks move the

disks at a rate of one disk per second, this number of moves would take 583,344,214,028 years, and thus it would take that many years for the world to end.

4. The optimal solution for the Towers of Hanoi with four pegs remains an open question.

Questions to Consider:

1. Here are the first few terms of a recurrence sequence. Can you find the rule that generates the next term using one or more previous terms?

    1, 1, 4, 13, 43, 185, …

2. Suppose there are 10 disks in your Towers of Hanoi puzzle. Use the method outlined in the lecture to compute how many moves are required to move all the disks to a new peg.

Lecture Six—Transcript
The Binet Formula and the Towers of Hanoi

Beyond their intrinsic appeal, and utility to number theorists, recurrence sequences are important objects in computer science. A question of considerable interest for mathematicians and computer scientists is, can we find a formula that will produce any individual term in a recurrence sequence of numbers without the need for generating *all* the numbers in the list up to that term?

In this lecture, we'll tackle this challenge by discovering the famous *Binet formula* for the Fibonacci numbers. While named after the French mathematician Jacques Binet, who first derived it in 1843, it appears that this important formula was known to Leonhard Euler and Daniel Bernoulli over 100 years earlier.

We'll verify this formula for ourselves by, once again, simply searching for a pattern. Once we derive this formula for ourselves, we'll realize that our method can be generalized and used to find corresponding formulas for all such similar recurrence sequences.

The Binet formula will provide us with the insight that while recurrence sequences such as the Fibonacci and Lucas numbers are not geometric progressions, they are in fact a combination of two geometric progressions. Yet again, we see structure.

We'll then close this lecture with one of the most famous stories involving recurrence sequences of numbers, "The Towers of Hanoi." While today these towers can be viewed as a challenging logic puzzle for mathematicians and computer scientists, as the ancient legend has it, the solution to this conundrum also holds the date at which the world will come to an end. Thus, we'll discover yet another useful piece of information that only comes into focus through the lens of number theory.

In the previous lecture, we discovered recurrence sequences: lists of numbers that can be generated using some starting seeds—the first few numbers on our list—and then the fixed rule for generating future numbers. Arithmetic and geometric progressions are very simple examples of recurrence sequences, as are the Fibonacci and Lucas numbers that we saw earlier.

A recurrence relation is one in which previous information is used in a systematic manner to generate new information. The concept of recurrence is important and fundamental—and actually an important component in many computer algorithms and languages. And as a result, computer scientists have studied and make use of recurrence sequences in their programming.

We now wonder if there is a so-called "closed formula" for the Fibonacci numbers. We begin by contrasting the idea of a recurrence definition and a formula.

First I want to name the Fibonacci numbers, which is something we haven't done yet. We've just listed them. Remember, they begin 1, 1, 2, 3, 5, 8, and so forth, but now I want to give them names in terms of their position in the list.

We name them by using the symbol F, with a subscript n, and this is read either "f sub n," for subscript n, or "f n" sometimes—it's abbreviated. What it means is the number n is denoting the location on the sequence at which we're at. For example, F_1 would denote the first Fibonacci number, or the Fibonacci number in the first spot, which would equal 1. F_2 would be the next one, the second Fibonacci number, which, remember, is also 1. F_3 would indicate the third Fibonacci number, which is 2; F_4 equals 3; F_5 equals 5, F_6 equals 8; and F_7, which means the seventh Fibonacci number on our list, would be 13; and so on. So I'm going to use F_n to denote the nth Fibonacci number, or a generic Fibonacci number.

This notation actually allows us to precisely define the recurrence rule that we've seen before. It would be said this way: $F_{n+1} = F_n + F_{n-1}$, and this holds for all natural numbers n greater than or equal to 2. What does this mean?

What it means is if you want to find the $(n + 1)^{th}$ Fibonacci number, what you do is add the two previous ones. Well, if this is the $(n + 1)^{th}$ Fibonacci number, what's the one right before it? It would be the nth Fibonacci number, and the one before that would be the $(n - 1)^{th}$. So $F_n + F_{n-1}$, if we add those two, we get F_{n+1}, and that's what that formula means.

However handy this notation might be, there is one disadvantage in defining the Fibonacci numbers as a recurrence sequence. That is that if we wish to find the value of a far-off one, say for example the 1000th Fibonacci number, which would be F_{1000}, we'd be required to compute all the previous Fibonacci numbers in succession and work our way up to the 1000th one, because the rule is a recurrence in the sense that we need the previous ones to get the next one. So we need to find all of them to get the 1000th one—all of them up to 1000.

Ideally, we would like a so-called "closed formula." In other words, a generic formula in which we can literally just plug in the number 1000 and produce the 1000th Fibonacci number without actually computing all the previous ones.

One of the important features of many recurrence sequences in general is that such closed formulas can be derived, and we'll illustrate this process with the famous Fibonacci numbers themselves.

We recall that in our attempt to express the golden ratio—which we denote by the Greek letter φ—in a more familiar form, rather than as a continued fraction of 1s, we saw that φ was one of the solutions to a particular quadratic equation, in particular: $x^2 = x + 1$.

Using the quadratic formula, we found that the two solutions to this equation are $(1 + \sqrt{5})/2$—and the other solution $(1 - \sqrt{5})/2$. Because the number φ—the golden ratio—is a positive number, we found that φ equals the positive one: $(1 + \sqrt{5})/2$. But now we're going to actually talk about the other solution as well, and so we'll give that a name. We'll use the Greek letter τ [tau] to denote that solution. In other words, τ will equal $(1 - \sqrt{5})/2$.

Here's an important insight. Since both φ and τ are solutions to the original quadratic equation, we see that if we plug those values in they will satisfy the equation. So, in particular, we see that $\varphi^2 = \varphi + 1$, and similarly, τ^2 must equal $\tau + 1$.

These important equations hold the key to finding a formula for the Fibonacci numbers, and I want us to see how we can use these identities to actually find the formula.

We can use the formula $\varphi^2 = \varphi + 1$ to simplify higher powers of φ. I want us to think about this together. Let's consider φ^3, which means $\varphi \times \varphi \times \varphi$. Well, φ^3 is nothing more than φ multiplied by φ^2. Now, φ^2, we know from the previous equation, is the same thing as $\varphi + 1$. So instead of having a φ^2, I can replace it by its twin, $\varphi + 1$. I have the $+ 1$ now, but notice no square. So I've reduced the complexity.

Let's come back to φ^3. Well, $\varphi^3 = \varphi \times \varphi^2$, and φ^2, in turn, is $\varphi + 1$. So $\varphi^3 = \varphi \times (\varphi + 1)$. If I distribute, I see that $\varphi^3 = \varphi^2 + \varphi$.

That's all well and good, but I have a φ^2, which again, I can simplify and reduce to just $\varphi + 1$. So I'd see $(\varphi + 1)$, plus that extra φ, or $2\varphi + 1$. In conclusion, we see that $\varphi^3 = 2\varphi + 1$. Lots of algebra, but if we work through it carefully, we see this identity holds.

Let's see if we can uncover a familiar pattern of numbers as we compute these higher and higher powers of φ. I'll just do a couple more just to illustrate the idea.

We can apply the exact same technique to find a simple way of writing φ to the 4th power. How? We proceed as we just did: $\varphi^4 = \varphi \times \varphi^3$; φ^3—we just discovered—was $2\varphi + 1$, so now I have $\varphi \times (2\varphi + 1)$. When I distribute that φ, I see $2\varphi^2 + \varphi$. Well, the φ^2—reverting back to our very first formula—is the same thing as $(\varphi + 1)$. So I now see 2 times the quantity $(\varphi + 1)$, plus an extra φ. So I have $2\varphi + \varphi$ is 3φ, and then I have plus 2.

So, now I see that $\varphi^4 = 3\varphi + 2$. Do you notice anything about these numbers in front of the φ and the number that we're adding? Well, again, it might not be clear. Let's look for the formula for φ to the 5th.

We continue: $\varphi^5 = \varphi \times \varphi^4$. And φ^4 we just found to be $(3\varphi + 2)$. So I have $\varphi \times (3\varphi + 2)$. When I distribute, I see $3\varphi^2 + 2\varphi$. Well, again, the φ^2 is nothing more than $\varphi + 1$. So I see $3 \times (\varphi + 1) + 2\varphi$; $3\varphi + 2\varphi$ is 5φ, and then I have that plus 3 at the end. Notice anything with respect to 5 and 3?

Finally, let's look at φ to the 6th. Let's just do one more. I promise; this is it: $\varphi^6 = \varphi \times \varphi^5$. I hope you see the pattern forming. That's φ times our previous answer, which is $5\varphi + 3$. When I distribute, I get $5\varphi^2 + 3\varphi$. Again I can see a φ^2, that's just $\varphi + 1$, all multiplied by 5. So 5φ plus 3φ is 8φ, and then I have a plus 5. Now, hopefully, a pattern is beginning to emerge.

If we collect all our findings, we'll see in a list:

$$\varphi^2 = \varphi + 1$$
$$\varphi^3 = 2\varphi + 1$$
$$\varphi^4 = 3\varphi + 2$$
$$\varphi^5 = 5\varphi + 3$$
$$\varphi^6 = 8\varphi + 5.$$

What do you see in that list? Notice that we see Fibonacci numbers. If you look down in the columns, you just see the Fibonacci numbers listed in order, and if you look across any particular row, you see two consecutive Fibonacci numbers: a big one followed by the smaller one that comes before it.

We see the Fibonacci numbers appearing. In fact, we could continue this process indefinitely and thus conclude that we have a general formula for any natural number n. So, here's the formula that we're seeing within the pattern. If I look at the nth power of φ, or φ^n, that will equal the nth Fibonacci number times φ, plus the Fibonacci number before that. Or, said mathematically: $\varphi^n = F_n\varphi + F_{n-1}$. Notice it's exactly what we just said. The nth power of φ equals the nth Fibonacci number times φ, plus the Fibonacci number that comes right before.

By the identical reasoning, since τ is also a solution of $x^2 = x + 1$, we can show that a corresponding formula holds for τ. In other words, for any natural numbers n, we also have that τ to the n equals the nth Fibonacci number times τ plus the Fibonacci number that comes before. Or said formulaic: $\tau^n = F_n\tau + F_{n-1}$.

If we take these two equations and now subtract the formulas, what do we see? Well, I see $\varphi^n - \tau^n$ on the left-hand side, and on the right-hand side, notice that the (F_{n-1})'s cancel when I subtract. So all I'm left with on the right-hand side is the nth Fibonacci number times the quantity $(\varphi - \tau)$. This is fantastic, because we can solve this for F_n, or the nth Fibonacci number.

All we have to do is divide both sides by $\varphi - \tau$, and I see the formula $F_n = (\varphi^n - \tau^n)$, all divided by $(\varphi - \tau)$. Now we have a formula for φ^n. We could simplify that a little bit more, because we can figure out what $(\varphi - \tau)$ is. What is it? Well, remember that φ is $(1 + \sqrt{5})/2$. Now we're going to subtract τ, which is $(1 - \sqrt{5})/2$, and so when we break this up, we see $1/2 + \sqrt{5}/2 - 1/2 - \sqrt{5}/2$. So the halves cancel, and I'm left with $\sqrt{5}/2 - (-\sqrt{5}/2)$, which means $\sqrt{5}/2 + \sqrt{5}/2$, which equals just $\sqrt{5}$. Thus we've derived a closed formula for the nth Fibonacci number in general: $F_n = (\varphi^n - \tau^n)/\sqrt{5}$. Or in other words, the nth Fibonacci number equals $(\varphi^n - \tau^n)/\sqrt{5}$. This elegant formula today is known as the *Binet formula*, named after the 19th-century French mathematician Jacques Binet.

By the way, we can now find the 1000th Fibonacci number. How do we do it? We just let n equal 1000. So the 1000th Fibonacci number equals $(\varphi^{1000} - \tau^{1000})/\sqrt{5}$.

I don't know about you, but I can't do that in my head—but a computer could simplify that, and it would produce a 209-digit Fibonacci number that begins 4346655… and so forth. Two hundred and nine digits. The important thing here is that we found this number without ever computing any of the previous Fibonacci numbers before it. This is the power of a closed formula. There was no need to work our way up. We landed right there. So, an absolutely powerful and beautiful identity.

Let's now recall from the previous lecture that we found that the Fibonacci sequence is not a geometric progression, since the ratio of consecutive terms is not constant. We did see that those ratios are converging to a particular value. The value the ratios are actually approaching, we saw, was the golden ratio: $(1 + \sqrt{5})/2$, also known as φ.

Using Binet's formula, we now discover that the Fibonacci numbers are, in fact, the difference of two geometric progressions. You see it? Because we have the $\varphi^n/\sqrt{5}$—that's a geometric progression, multiplying by φ each time, and then we subtract off $\tau^n/\sqrt{5}$, where again, a geometric progression, we multiply by τ each time. So it's absolutely wonderful that even though we see that the Fibonacci numbers are not examples of a geometric progression, we see that since those ratios of terms were converging to something, something else was happening: It's actually the difference of two geometric progressions. Again, we see structure.

We could actually apply the previous analysis exactly with the Lucas sequence, and we'd derive a different closed formula. That closed formula would be: For all natural numbers n greater than or equal to 2, the nth Lucas number would be equal to $\varphi^{n-1} + \tau^{n-1}$. Notice it's a similar-looking formula; it involves both the φ and the τ. In this case we're adding and we have different exponents.

In fact, using the same ideas, the terms of any such recurrence sequence can be expressed as a closed formula.

I want to close this lecture on recurrence with the legend of the Towers of Hanoi. The Towers of Hanoi was a logic puzzle that was marketed in 1883 by a so-called "Professor Claus." However, the name "Professor Claus" is, in fact, an anagram of its true inventor, Professor Lucas. Maybe he invented this game to subsidize his retirement. I don't know. But anyway, he was marketing this thing, which is kind of neat.

The Towers of Hanoi puzzle consists of three pegs, and I actually have one here. So this is what it looks like. Usually they're made out of wood. It consists of three pegs and a collection of punctured disks. You can see these punctured disks here. They have holes in them so they fit over the pegs, but the disks are of different sizes.

The puzzle begins with all the disks on one peg in descending order. So, we go from the largest on the bottom, up to the smallest on the top. The object is to transfer all the disks to another peg so that they end up residing on this new peg in the original descending order. What we want to do is we start this way, and we want to end, say, with all of them sitting here, but there are some rules, because otherwise the solution would be literally what I'm doing right now. So there are some rules, and here are the rules.

First of all, only one disk can be moved at a time. So only one disk can be moved at a time, and at no time can a larger disk be placed on top of a smaller disk. So while I could take this disk and put it here, I would not be permitted to put it here. That's against the rules because I have a larger one on top of a smaller one. That's not allowed. So that is against the rules.

The challenge is to find a method for moving the disks and to determine the smallest number of moves required to solve the puzzle—to move all of them. This is truly a towering recurrence, as we're about to see.

Let's suppose that in general we have n disks to be moved, for some natural number n. We begin by observing that to move the largest, very bottom disk, we must first remove all the other $n - 1$ smaller disks.

Once we remove those $n-1$ disks to a new peg, then we can actually free up that last, biggest disk to be moved onto the empty peg. However, now we must remove all the previous disks and move them back on top of the largest one. Let's now give this a name—the number of steps involved in solving this—a name. Let's write h_n for the number of moves required to transfer n disks. In other words, h_1 would equal the number of moves required to move one disk, and so forth.

We now wonder if we can find a recurrence rule that actually generates the sequence of numbers h_n. Let's see if we can find a pattern by coming back to the puzzle. Instead of performing the puzzle with all these disks, let me just start by looking at 3 disks first. How many moves are required, which would be equal to h_3? I have 3 disks—n would be 3 here—h_3 would be the minimum number of moves.

If we try it, we see what happens. I'll move this smallest one to here, then I'll move the middle one to this empty one. Now to move this largest one, I can't put it on top of either of these, so I'll take the little one and put it on top of this, which is allowed, because the small is on top of the medium. Then the large is now freed up to be moved to here, and then now I can't pick these both up, but I can now take the little one and put it on this empty one, take the medium one, put it back on the big one, and the little one gets back on top.

How many moves was that? I'll do it a little bit quicker now, and now all we'll do is just count. So if we count, we see 1, 2, 3, 4, 5, 6, 7. Seven moves are required, and in fact that's the shortest, the fewest number of moves required to do 3 disks.

Let's now consider 4, so I'm going to add on a bigger one. To do 4, let's just think about it. I first have to move all of these off, then move this, and then move them all back on. So let's see what happens if we do this. I'm going to count right now, so here we go live: 1, 2, 3, 4, 5, 6, 7. So there was the 7 to move it. Now this is free, so I pick this up, and I'm going to put it here. So I had 7. Now I just moved it, that's an 8th move, and now we've got to continue. So an 8th move, 9, 10, 11, 12, 13, 14, 15. *Voilà*. I did it.

Amazing as it may seem, I was able to do this, which means that if we look at $n = 4$, the answer is 15. So, h_4 is 15. Let's think about what's going on here. In view of this disk-moving process, h_n actually must equal $h_{n-1} \ldots$. First move the top $n-1$ disks, then you move that biggest disk, which is another move, so plus 1 move; and then you've got to move all these from where they are back on top; so that's another h_{n-1}.

We see that h_n has to equal $h_{n-1} + 1 + h_{n-1}$ again. In other words, we have a recurrence rule: $h_n = 2h_{n-1} + 1$. Remember that moving that bottom disk is where we get the plus 1.

So, the number of moves required for 4 disks, if we use this formula, equals 2 times the number of moves required to move 3, which we saw was 7, so $2 \times 7 + 1$, which is 15, which is what we got when we worked it out for ourselves. So this is actually good.

The question is, can we find a closed formula by looking at our sequence of h's? So we have a recurrence definition for h. Now we want to see if we can find a closed formula for h_n. Let's look at the numbers. One, so if we just had 1 disk, it's easy: just literally move it and so 1. Then the next count would be 3, the next count is 7, which is what we saw, and then 15. Do we notice anything about 1, 3, 7, 15?

Again, we're always looking for a pattern. In this case, we notice that each number is actually 1 less than a power of 2. Let's check it out for ourselves. Notice that 1 equals $2 - 1$; 3 equals $4 - 1$, which is $2^2 - 1$; 7 equals $8 - 1$, which is $2^3 - 1$; and even 15 is equal to $16 - 1$, which is $2^4 - 1$.

In general it's actually possible to prove that h_n—the number of moves required with n disks—is equal to $2^n - 1$, which gives our simple closed formula for the number of disk moves required if we have n disks.

For our model with 7 disks, you might say, "Gee,"—in fact, you might be a little bit disappointed that I'm doing it only with 4 and I didn't revert back to the original one, which, notice, has 3 more. So it actually has a total of 7 disks—"Gee, how come you didn't do it with 7 disks?" Well, we can now see how many moves that would take.

It would equal $2^7 - 1$. The closed formula gives it to us immediately, which is $128 - 1$, or 127 moves. So if you want to solve this puzzle, you'll need a little bit of time because it's going to require 127 moves.

Now using this formula, we can actually determine the end of the world. There's a legend that a certain group of monks had a particularly impressive edition of this puzzle consisting of 64 gold disks and three diamond pegs. They were able to move—the monks were able to move these disks at a pretty alarming rate: 1 disk per second.

The legend is that the world would end once the monks completed their mission. We can now use the closed formula we just found to predict when the world will end according to legend. The number of moves required to solve this puzzle with 64 disks equals $2^{64} - 1$. You see the power of the closed formula. Given that the monks move the disks at a rate of 1 disk per second, this number of moves would take 583,344,214,028 years, and thus that many years for the world to end. So you might think about putting a hold on your mail.

Anyway, what if we had four pegs? Let's go back to the puzzle and add an extra degree of freedom—namely, add an extra peg here, which I don't have at our disposal, so let's just use our imagination and see what would happen if we had four pegs.

Well, if we had a fourth peg, then in fact instead of taking 7 steps we would actually require fewer. Let's try it with just three. We used 7 before. But now what I'll do is I'll take this little one and put it on the invisible fourth peg. Then I can just move this to here, move the big one to here, put the medium back to here, and then put this back to here.

How many moves was that? That was actually only 5 moves. So we actually, with that extra peg, we'd have an extra degree of freedom, and so therefore we'd need fewer moves. Certainly less than 7.

It turns out that the optimal strategy and the optimal solution for the Towers of Hanoi with four pegs actually remains an open question. So you can see just adding one peg leads us from the world that we understand and drives us into the world of the unknown. Open questions abound.

The importance here is that we see that recurrence sequences have a particularly nice form and certain recurrence sequences can actually lead to a closed formula—a way of generating a far, far-out term, without going through the steps of finding all the terms before. Amazing number theoretic structure within recurrence sequences.

Lecture Seven
The Classical Theory of Prime Numbers

Scope:

Here we introduce the ideas that feed analytic number theory—the study of prime numbers. The main goal of this classical area of study is to uncover the distinctive personalities of natural numbers through the arithmetic structure that unfolds from multiplicative considerations—specifically, from expressing natural numbers as products of the smallest possible factors greater than 1. This factorization idea will allow us to partition the natural numbers greater than 1 into two disjoint collections—the prime numbers and the composite numbers. We will discover why 1 is neither prime nor composite—an issue that will also foreshadow the birth of algebraic number theory. In this lecture we will discover the 2000-year-old struggle to understand the primes that started in ancient Greece with important contributions by Euclid and Eratosthenes. Along the way, we will encounter the fundamental theorem of arithmetic and establish this important result through a "divide and conquer" argument. We will also see how to create a "sieve" to sift out the primes from the composite numbers and then discover that there are, in fact, infinitely many prime numbers. Euclid established this result, which is considered to be one of the most elegant proofs in mathematics. Finally, armed with the reality that there are infinitely many primes, we wonder if the prime numbers appear with any regularity within the natural numbers.

Outline

I. The story of the prime numbers.

   A. The basic building blocks of number theory.

      1. How can we generate the natural numbers? There are several different methods involving addition.

      2. The simplest is to start with 1 and continue to add 1 repeatedly. In fact, this is the simplest example of an arithmetic progression.

      3. If we start with the triangular numbers as building blocks, then we could apply Gauss's profound result that

he discovered at the age of 19: Every natural number is the sum of at most three triangular numbers.

4. Alternatively, in our discussion of Fibonacci numbers we outlined a method for converting kilometers to miles. Implicit in that method is the fact that every natural number can be written as the sum of distinct Fibonacci numbers.

5. Here we will consider the basic multiplicative building blocks of the natural numbers.

B. Factorization as a personality trait of numbers.

1. To understand the arithmetic structure and individuality of the natural numbers better, we study their basic components when viewed as products of smaller numbers.

2. The process of expressing natural numbers as products of smaller numbers is known as *factoring*. The terms in the product are called *factors*.

3. The individual factors reveal features of the number. For example, if a number has a factor of 2, then it must be even. If a number has a factor of 10, then the number's last digit must be 0.

C. A brief history of the primes.

1. We define the prime numbers to be the atoms of the natural numbers—those that cannot be split into smaller pieces. More precisely, a natural number greater than 1 is called *prime* if it cannot be expressed as the product of two smaller natural numbers.

2. The first few primes are 2, 3, 5, 7, 11, 13, 17, 19, and 23. The number 6 is not prime because it can be written as 2 × 3. Similarly, 15 is not prime since it equals 3 × 5. Numbers that are greater than 1 and not prime are called *composite numbers*.

3. Euclid, around 300 B.C.E., was the first to embark upon a rigorous and formal study of the prime numbers and some of the properties they possess.

II. The unique factorization property.

A. Euclid's *Elements*.

1. Euclid's *Elements of Geometry*—a series of 13 books many consider to be some of the most important treatises

ever written—contain approximately 100 important
results involving number theory.

2. Two of his results—which we will now explore—are the
cornerstones for the two main branches of number
theory.

B. A unique factorization property.

1. The first theorem of Euclid that we highlight here is
what is now known as the *fundamental theorem of
arithmetic*, which in essence asserts that the primes are
indeed the multiplicative building blocks of all natural
numbers greater than 1.

2. The fundamental theorem of arithmetic: Every natural
number greater than 1 can be expressed uniquely as a
product of prime numbers.

3. For example, 12 can be expressed as $2 \times 2 \times 3$, and
except for rearrangement of these prime factors, this
prime factorization is unique. Thirteen is a prime so we
would write it as the "product" 13, again in only one
way.

C. A "divide and conquer" argument.

1. The argument that established the validity of the
fundamental theorem of arithmetic involves the
technique of "divide and conquer."

2. Suppose we are given a natural number $n > 1$. If it is a
prime number, then we have factored it into primes and
we are done. If n is not a prime (so it is composite), then
by definition it can be expressed as the product of two
smaller natural numbers.

3. We now repeat this process with each of the two smaller
factors. This process will eventually terminate; that is,
we will not be able to factor the numbers any further. In
other words, we are left only with prime factors, as
desired.

4. Verifying the uniqueness of the factorization into primes
is much more subtle, although intuitively it seems
reasonable. We will return to this uniqueness of
factorization into primes in our excursion into algebraic
number theory, in particular, in Lecture Fifteen.

5. This uniqueness property also helps us understand why
we do not consider 1 a prime number. If it were prime,

then we would not have unique factorization; for example, $6 = 2 \times 3 = 1 \times 2 \times 3$.

III. The Sieve of Eratosthenes.

    A. Eratosthenes and his work.

        1. Given that the primes are the fundamental multiplicative building blocks for the natural numbers, early on in the human history of number theory there was a desire to identify which numbers are primes.

        2. Around 200 B.C.E., Eratosthenes discovered a method for taking the natural numbers and "sifting" out the composite numbers. Thus his "sieve," now known as the *Sieve of Eratosthenes*, collected all the prime numbers up to any particular value.

    B. Sifting out the composite numbers.

        1. Suppose we wish to list all the primes less than 100. Eratosthenes's method is to write all the numbers from 2 through 100. We start at the first number, 2, and from there we cross every other number off our list. In this case, we would cross off 4, 6, 8, 10, and so forth; all the even numbers, except 2, would be crossed off.

        2. We move to the next number not crossed off our list, in this case, 3. From there we cross off every third number (with the understanding that we might be crossing off numbers that have already been removed). In this case, we remove all the multiples of 3: 6, 9, 12, 15, 18, and so forth.

        3. If we repeat this process, then when we are finished, the only numbers *not* crossed off are precisely all the prime numbers less than 100.

        4. It might appear as if there are a large number of steps required to get all these primes. However, Eratosthenes showed that we need only repeat this pruning process up to the square root of 100, which is 10. That is, we need only perform four steps (one for each prime less than 10: 2, 3, 5, and 7) to generate *all* the primes up to 100!

C. Why this sieve works so quickly.

   1. Initially it seems surprising that we need only four steps to generate all the primes up to 100.

   2. We need to prove that after we sieve out by all the numbers up to 10, the numbers beyond 10 that have *not* been sifted out are all primes. We will establish this assertion by assuming the opposite and producing a contradiction.

   3. Suppose that there were a composite not crossed off our list. Then it is not a multiple of any number less than or equal to 10, so it is the product of two numbers greater than 10. Thus it is greater than 10 × 10, or 100, which is too large to appear on our list.

   4. Therefore we see that there cannot be any composite numbers left beyond 10.

   5. In general, if we wish to find all the primes up to *n*, we need only sieve up to $\sqrt{n}$.

D. A larger example.

   1. Suppose we wanted to list all the primes less than 500.

   2. This at first appears to be a daunting task. However we note that $\sqrt{500} = 22.3606\ldots$, so all we need to do is list the natural numbers from 2 to 500 and repeat this sieving process until we reach 22.

   3. It is easy to check that the complete list of primes less than 22 is 2, 3, 5, 7, 11, 13, 17, and 19. Thus just after eight sifting processes, we will have generated *all* the primes up to 500!

IV. How many primes are there?

A. Euclid's result on the infinitude of primes.

   1. Inspired by Eratosthenes's sieve, we now wonder how many primes there are.

   2. This brings us to the second theorem of Euclid that we celebrate in this lecture. In Book IX of his *Elements of Geometry*, Proposition 20 states: Prime numbers are more than any assigned multitude of prime numbers.

   3. Today we would state this assertion as: There are infinitely many prime numbers.

   4. This extremely important 2300-year-old result of Euclid's is the pillar upon which analytic number theory was built. We will study the later refinements of this theorem in the next lecture.

5. However, here we will follow Euclid's ingenious proof that establishes this great theorem.

B. Searching for a prime greater than 3.
   1. To build our intuition, we first begin simply: How can we find a prime beyond 2 and 3?
   2. One method is to announce, "5," and move on, but this method does not easily generalize.
   3. Another approach is to first consider the number 2×3. Of course this number is certainly not prime, since it is the product of both 2 and 3. However Euclid's clever idea is to add 1; that is, he considered the number $2 \times 3 + 1$ and argued that any prime dividing this number cannot be 2 or 3.
   4. Thus Euclid proved there exists a prime beyond 3 without using the fact that 3 is a specific small prime. Euclid's devilishly clever idea can be extended to prove his theorem in general.

C. Euclid's beautiful proof.
   1. Let p_1, p_2, p_3, …, p_n be the first n prime numbers. Our goal is to prove that there must exist another prime not on this list.
   2. We remark that each of the first n prime numbers divides evenly into the product of all of them: $p_1 \times p_2 \times p_3 \times \cdots \times p_n$.
   3. Euclid then considers the number $E = p_1 \times p_2 \times p_3 \times \cdots \times p_n + 1$. Now, E is a natural number greater than 1, so we know by the fundamental theorem of arithmetic that it can be factored into primes. Let us say that q is one of the prime numbers that divide evenly into E.
   4. Can the prime q be one of the first n prime numbers? Since E is the product of the first n primes plus 1, we see that none of these n primes can divide evenly into E. When we divide any of those primes into E, we find a remainder of 1.
   5. Thus q must be a prime number that is not contained in the first n primes. Therefore we conclude that there are more than n primes—and since n was an arbitrary number, there must be infinitely many primes.

6. Most mathematicians view Euclid's proof as one of the most elegant arguments in all of mathematics.

V. Are there long runs of composite numbers?

  A. Runs of composite numbers.

    1. We have seen that there are infinitely many prime numbers and infinitely many composite numbers.

    2. We cannot have two consecutive numbers both being prime beyond 2 and 3, since every other number is even (and thus has a factor of 2).

    3. Can we find two consecutive numbers both of which are composite? Yes: 8 and 9 is the smallest such pair. In fact, 8, 9, 10 is a run of three consecutive composite numbers. We note that 24, 25, 26, 27, and 28 is a run of five consecutive composite numbers.

  B. What is the maximum run of consecutive composite numbers before hitting a prime?

    1. It is a theorem that there are arbitrarily large runs of composite numbers. That is, given any natural number N, there exists a run of N consecutive composite numbers.

    2. Phrased differently, given any natural number N, two prime numbers exist whose distance from each other is at least N and for which there are no other prime numbers between them.

    3. More informally, we can find arbitrarily long runs of natural numbers that are deserts free of primes—there are none to be found!

  C. A modification of Euclid's argument.

    1. To prove the assertion that given any natural number N there exists a run of N consecutive composite numbers, we invoke a clever modification of Euclid's method of proving there are infinitely many primes.

    2. We start with the number K, defined to be the product of the natural numbers given by $K = 2 \times 3 \times 4 \times \cdots \times (N + 1)$. We observe that K is divisible by all the numbers from 2 to $N + 1$. (Note that there are N numbers in this list of divisors.)

3. We now consider $K + 2 = 2 \times 3 \times 4 \times \cdots \times (N + 1) + 2$. The number $K + 2$ is divisible by 2 and therefore is composite.

4. The next number, $K + 3$, equals $2 \times 3 \times 4 \times \cdots \times (N + 1) + 3$, and we note that this number is also composite, because 3 is a factor of $K + 3$. Similarly, we see that 4 is a factor of $K + 4$, so it is composite.

5. This pattern continues all the way through the number $K + (N + 1)$, which is composite because $N + 1$ is a factor of $2 \times 3 \times 4 \times \cdots \times (N + 1) + (N + 1)$.

6. Thus we see that each of the N consecutive numbers, $K + 2$, $K + 3$, ... , and finally $K + (N + 1)$, is composite. So we found a run of at least N consecutive composite numbers.

7. This argument is effective in that it gives a method for finding the run of numbers, although it is not practical for long runs.

8. By creating a variation on Euclid's theme, we were able to produce an entirely new result.

Questions to Consider:

1. Using the method of Euclid and the prime numbers 2, 3, 5, 7, 11, 13, 17, and 19, explicitly describe a natural number that contains a prime factor greater than 20.

2. Describe a number that is itself composite and for which the next 1 million consecutive natural numbers are all composite numbers.

Lecture Seven—Transcript
The Classical Theory of Prime Numbers

In this lecture, we introduce the ideas that feed analytic number theory—the study of prime numbers. The main goal of this classical area of study is to uncover the distinctive personalities of the natural numbers through the arithmetical structure that unfolds from multiplicative considerations, specifically, from expressing natural numbers as products of the smallest possible factors greater than 1.

This factorization idea will allow us to partition the natural numbers greater than 1 into two distinctive collections: the *prime numbers* and the *composite numbers*. We'll also discover why the number 1 is neither prime nor composite—an issue that actually foreshadows the birth of algebraic number theory, an area that we'll investigate later in this course.

Here we'll discover the 2000-year-old struggle to understand the primes that started in ancient Greece with important contributions by Euclid and Eratosthenes. Along the way, we'll encounter the fundamental theorem of arithmetic, and establish this important result through a "divide and conquer" argument.

We'll also see how to create a mathematical "sieve" to sift out the primes from the composite numbers and then discover that there are, in fact, infinitely many prime numbers. The reasoning that establishes this result, due to Euclid, is considered to be one of the most elegant proofs in all of mathematics.

Finally, armed with the reality that there are infinitely many primes, we'll wonder if the prime numbers appear with any regularity within the natural numbers.

This lecture introduces the basic building blocks of number theory. So, we start by asking, how can we generate the natural numbers? There are several different methods involving addition. The simplest is to start with 1 and continue to add 1 repeatedly. In fact, this is the simplest example of an arithmetic progression, which we studied earlier.

If we start with the triangular numbers as building blocks, then we could apply Gauss's profound result that he discovered at the age of 19 that states that every natural number is the sum of at most three triangular numbers. If we begin with all the triangular numbers and then just start adding together the numbers in various combinations, eventually we'll generate all the natural numbers.

Alternatively, in our discussion of the Fibonacci numbers, we actually outlined a method for converting kilometers to miles. Implicit in that method is the fact that every natural number can be written as the sum of distinct Fibonacci numbers—yet another way of generating natural numbers using addition.

But now, rather than using addition, here we'll consider the basic multiplicative building blocks of the natural numbers.

In order to better understand the arithmetical structure and individuality of the natural numbers, we study their basic components when viewed as products of smaller numbers. The process of expressing natural numbers as products of smaller numbers is known as *factoring*. The terms in the products are, in fact, called *factors*.

The individual factors reveal features about the number. For example, if a number has a factor of 2, then it must be even. If a number has a factor of 10, then the number's last digit must be 0. Again, we see insights into the number by looking at its factors.

We now define the prime numbers to be the atoms of the natural numbers—those that cannot be split into smaller pieces. More precisely and more rigorously, a natural number greater than 1 is called *prime* if it cannot be expressed as the product of two smaller natural numbers.

The first few primes are 2, 3, 5, 7, 11, 13, 17, 19, 23, and so on. Notice that the number 6 is not a prime, because it can be written as 2 × 3—the product of two smaller numbers. Similarly, 15 is not prime since it can be written as 3 × 5—again, two smaller natural numbers. Numbers that are greater than 1 and not prime are called *composite numbers* because they are composed of smaller numbers.

Euclid, around 300 B.C.E., was the first to embark upon a rigorous and formal study of the prime numbers and the properties that they in fact possess. Euclid was the author of a series of 13 books entitled *The Elements of Geometry*, and many scholars today consider these works to be some of the most important treatises ever written. The texts contain approximately 100 important results involving number theory.

Two of his results, which we'll now explore in this lecture, are the cornerstones for the two main branches of number theory: analytic number theory, which we'll discuss in here; and algebraic number theory, which we'll discuss later in the course.

The first theorem of Euclid that we highlight here is what is now known as the *fundamental theorem of arithmetic*, which in essence asserts that the primes are indeed the multiplicative building blocks of all natural numbers greater than 1.

The fundamental theorem of arithmetic asserts that every natural number greater than 1 can be expressed uniquely as a product of prime numbers. For example, 12 can be expressed as $2 \times 2 \times 3$—and notice those are all prime numbers. Except for the rearrangement of those factors, these primes are in fact unique. So, we have this unique factorization. I could write $2 \times 2 \times 3$ or $2 \times 3 \times 2$ or $3 \times 2 \times 2$, but in all those factorizations, we always have the two occurrences of the prime 2 and the one occurrence of the prime 3. So in that sense, it's unique.

What about the number 13? Well, 13 is prime, so we would consider it as its own little product, just 13, left alone; and again, there's only one way to write 13 in this fashion.

The argument that established the validity of the fundamental theorem of arithmetic involves a technique we might informally call "divide and conquer," which we've seen before. Here's the idea of the proof using the number 90.

Since 90 is not a prime number, we can factor it into two smaller natural numbers, say, $90 = 9 \times 10$. We now look at each individual factor, so we look at 9. Well, is 9 a prime? No. So it can be factored. In fact, $9 = 3 \times 3$. But notice that 3 is a prime, so we can't break down those 3s any further.

We now come back to the other factor, 10. What can we say about that? That's not prime, so we can "divide and conquer" it—write it as 2×5. Notice that again, each of these factors are, indeed, primes. Therefore, we see that $90 = 2 \times 3 \times 3 \times 5$. Notice how we divided and conquered—we took a big number and kept breaking it apart.

To begin the proof of the theorem in general, let's suppose that we're given a natural number n, let's call it, which is greater than 1. If it is a prime number, then we factored it into primes, namely, just itself, and so we're done.

Now let's consider the other case. If n is not a prime number—so in other words, it's a composite number—then by definition it can be expressed as the product of two smaller natural numbers. Here's the dividing and conquering coming in. Well, we now repeat this factoring process with each of the two smaller factors, and this process will eventually terminate. In other words, we'll not be able to factor the numbers from some point on any further because those numbers would have been prime factors, as desired.

Verifying the uniqueness of the factorization into primes is much more subtle, although intuitively it seems reasonable when we look at examples. We'll return to this uniqueness of factorization into primes in our excursion into the other main branch of number theory: algebraic number theory. In particular, we'll revisit this issue in Lecture Fifteen.

This uniqueness property also helps us understand why we don't consider 1 to be a prime number. Let's imagine for a moment that we considered 1 to be a prime. Then we would lose the unique factorization feature. For example, 6 could be on the one hand 2×3—a product of primes—but on the other hand, it could be $1 \times 2 \times 3$, which would be a different product of primes, if we were allowed to count 1 as a prime. Therefore, we don't consider 1 to be a prime.

We now face the question, how can we find the prime numbers? Given that the prime numbers are fundamental multiplicative building blocks for the natural numbers, early on in the history of number theory there was a desire to identify which numbers are in fact prime. Around 200 B.C.E., Eratosthenes of Alexandria discovered a method for taking the natural numbers and, in some sense, "sifting" out the composite numbers. Thus, his "sieve"—as it's now referred to—is called the *Sieve of Eratosthenes*. In some sense, the sieve collected all the prime numbers up to any particular given value.

How did his process actually work in practice? Let's consider an example. Suppose we wish to list all the primes less than 100. In Eratosthenes's method, what we do is we write the numbers from 2 through 100. We then start with the first number, in this case 2, and from there we cross off every other number. In this case, we would cross off the numbers 4, 6, 8, 10, and so forth. All the even numbers after 2 would be crossed off—that is, we'd be eliminating all the multiples of 2 that are larger than 2.

We then move to the next number not crossed off our list—in this case we quickly come upon 3. From there, we cross off every third number, with the understanding that we might be actually crossing off numbers that have already been removed. In this case, we'd remove all the multiples of 3 that are larger than 3, in particular, 6, 9, 12, 15, 18, and so forth.

If we repeat this process, the next number not crossed off our list would be 5. You might say, "What about 4?" Notice that 4 was crossed off our list since it was already seen to be a multiple of 2. We next cross off every fifth number after 5, thus removing all multiples of 5, and so on. When we're finished, the only numbers *not* crossed off are precisely all the prime numbers less than 100.

It might appear as if there's a large number of steps required to find these primes. However, Eratosthenes showed that we need only repeat this pruning process up to the square root of 100. The square root of 100 equals 10, because 10×10 gives us 100.

In other words, to generate all the primes up to 100, we need only perform four steps, one for each of the primes less than 10, in particular, 2, 3, 5, and 7. So, in four steps, we produce *all* the primes up to 100.

Why does this sieve work so quickly? Intuitively, it seems surprising that we need only four steps to generate all the primes up to 100, so let's think about it and retrain our intuition. We need only prove that after we sieve out by all the numbers up to 10, the numbers beyond 10 that have *not* been sifted out are all primes.

We'll establish this assertion by assuming the opposite and producing a contradiction. So we'll consider a proof by contradiction. Suppose contrary to what we hope, or what we claim, that there was a composite number not crossed off our list. Then it is not a multiple of any number less than or equal to 10. So, it's the product of two numbers, each greater than 10, thus the number itself must be greater than 10×10, which equals 100, which is too large to appear on our list of the first 100 numbers. Therefore, after we've sieved out all the prime numbers less than 10, we see that there cannot be any composite numbers left on our list.

In general, if we wish to find the primes up to, let's say a number n, where n is some arbitrary natural number, we need only sieve up to the square root of n.

To illustrate the power of this result, let's consider a larger example. Suppose we wanted to list all the primes less than 500. Now that's a lot. At first it appears that this is really genuinely a daunting task. However, we can use a calculator to compute the square root of 500 and see that it's 22.36-something. So all we need to do is to list all the natural numbers from 2 to 500, which we could write out—it would take a while, but we could write them all out—and then repeat this sieving process until we reach the number 22.

It's easy to see that the complete list of primes less than 22 is 2, 3, 5, 7, 11, 13, 17, and 19. Thus, after just eight sifting processes, we'll have generated all the prime numbers up to 500. Truly remarkable.

Inspired by Eratosthenes's sieve, we now wonder how many primes there are altogether. This brings us to the second great theorem of Euclid that we celebrate in this lecture. In Book IX of his *Elements of Geometry*, Proposition 20 stated that the prime numbers are more than any assigned [multitude] of prime numbers. Kind of a mouthful, so let's play that back and see what it says.

The prime numbers are more than any assigned magnitude, so let's assign a magnitude to the primes. Suppose that there were 10 primes. This theorem asserts that the primes are more than that magnitude. Okay. That means there's more than 10 primes. What if we consider the magnitude of 100? Well, this theorem, this proposition asserts that there will be more primes than that magnitude, so there will be more than 100 primes.

Today we would state this assertion a little bit differently; we'd state it simply as, "There are infinitely many prime numbers." This extremely important 2300-year-old result of Euclid is the pillar upon which analytic number theory was built. We'll study later refinements of this fundamental result in the next lecture. But here, however, we'll follow Euclid's ingenious proof that establishes this great theorem. So we'll actually walk in Euclid's sandals right now.

To build up our intuition, we first, as always, begin simply. So let's answer the following question: How could we find a prime beyond 2 and 3? One method is to simply announce, "5," and move on. That's great, except that that method doesn't really easily generalize if I want to find, let's say, a prime after a billion.

Another approach is to first consider the number 2×3. Of course, this number is certainly not prime, since it's the product of both 2 and 3. However, Euclid's clever idea is to add 1. In other words, he considered the number $2 \times 3 + 1$ and argued that any prime number dividing this number cannot be 2 or 3. Why is that?

We can compute $2 \times 3 + 1$ and see 7. But don't think about it in terms of 7. Think about it as $2 \times 3 + 1$. Notice that 2 is a factor of that first number in the sum, and so when I add the 1, I'd see I'd have a remainder of 1 when I divide by 2. So, 2 doesn't evenly divide this number. Similarly, since 3 divides 2×3 evenly, then when I add the 1, I'd see a remainder of 1 when I divide this number by 3. So neither 2 nor 3 can divide this number evenly.

By the fundamental theorem of arithmetic, we know that there has to exist prime numbers that evenly divide $2 \times 3 + 1$, because every natural number bigger than 1 can be factored into primes. So there's got to be some prime out there that divides this number, but we see it can't be 2 and it can't be 3. Therefore, there has to be a prime greater, which in this case is the number itself, 7.

Thus, Euclid proved that there exists a prime beyond 3 without ever using the fact that 3 is a specific small prime. Euclid's devilishly clever idea can be extended to prove his theorem in general. So let's have a look. In fact, let's now celebrate Euclid's beautiful proof.

I'm going to name the prime numbers up to a certain point. Let's name the first n prime numbers. Of course, we know they start with 2, 3, 5, 7, and so forth, but let me give them a little bit more generic names so we can talk about them systematically. I'll call p_1 the first prime, so that's 2; p_2 the second prime, so that's 3; p_3 the third prime, which is 5; but I'll keep listing them all the way out to p_n. So p_n is the nth prime. And even if n is really, really large and we don't happen to know the nth prime, p_n will represent that number.

So p_1, p_2, out to p_n, will be the first n prime numbers. Our goal is to prove that there must exist another prime number that's not on this list. We remark that each of the first n prime numbers divides evenly into the product of all of them, $p_1 \times p_2 \times p_3$, all the way out to multiply by p_n.

Euclid then considers the auxiliary number, which I'll call E, which is the product of all those primes, up to p_n, and then we add 1. E is a natural number greater than 1, so we know by the fundamental theorem of arithmetic that it can be factored into primes. Let's say that q is one of the prime numbers that divides evenly into this auxiliary number, E.

Can the prime number q be one of the first n prime numbers? Since E is the product of the first n primes, with a "+ 1" at the very end, we see that none of these n primes can divide evenly into the number E. When we divide by any of these primes into E, when we divide any of those primes into E, we'd find a remainder of 1. So they don't divide.

Thus, q must be a prime number that is not contained in the first n primes. Therefore, we conclude that there must be more than n primes, and since n was an arbitrary number—it could be anything— there must be, in fact, infinitely many primes. Because whenever you think you've got them all, I just run this argument with that as n and I find a prime that's beyond that list.

Most mathematicians view Euclid's proof as one of the most elegant arguments in all of mathematics, and why? Well, because it was the incredibly insightful and powerful, profound idea of constructing this auxiliary number to be the product of all the primes up to a point and then adding 1.

Once we're handed that number, if we analyze the proof and think about it on our own, we can actually see that the rest of the result follows suit—the proof now follows easily. But it was this incredibly delicate idea of constructing this number, which on one hand is both subtle and on the other hand simple, that makes this argument so attractive and so aesthetically appealing.

We close this lecture by wondering if the prime numbers appear with any regularity within the natural numbers. We've seen that there are infinitely many prime numbers and, of course, infinitely many composite numbers—numbers that are not prime. Notice that we cannot have two consecutive numbers both being prime beyond 2 and 3, since every other number is even—in other words, has a factor of 2 in it.

Can we find two consecutive numbers both of which are composite? The answer is yes. For example, 8 and 9, notice, are in fact the smallest such pair—two numbers, each of which are composite and they're consecutive. In fact, 8, 9, 10 is actually a run of three consecutive composite numbers. Moreover, we note that 24, 25, 26, 27, and 28 is actually a run of five consecutive numbers that are each composite. So there's a run of five composite numbers that are consecutive.

What's the maximum run of consecutive composite numbers that we could have before we hit a prime? It turns out that it's a theorem that there are arbitrarily long runs of composite numbers. In other words, given any natural number N, there exists a run of N consecutive composite numbers, or phrased differently, given any natural number N, there exists two prime numbers whose distance from each other is at least N and for which there's no other prime numbers in between them.

More informally, we can find arbitrarily long runs of natural numbers that are deserts free of primes—there are no primes to be found. In some sense, we're walking along the natural numbers and we'll come upon these places, these deserts where we're thirsting for a prime, and yet there's no prime to be found. We know if we walk long enough we will eventually hit one, but that journey might be very, very long. In fact, as long as we wish.

To prove this assertion that given any natural number there exists a run of N consecutive composite numbers, we invoke a clever modification of Euclid's method of proving that there are infinitely many primes. Thus, we're about to see a wonderful illustration of how we should always try to extend great ideas whenever and wherever we find them—a great life lesson for us.

To inspire the argument, let's find two consecutive composite numbers. We've already seen that 8 and 9 are such an example, but now we're looking for a process that can be extended for a run of consecutive composite numbers that could be anything. So let's return to 2×3 in the inspiration for Euclid's argument. But now let's add 2, so $2 \times 3 + 2$.

Notice that this number is composite because it's divisible by 2. Notice that 2 divides the product 2×3, and 2 divides that 2 all by itself. So therefore, 2 divides the number.

What's the next number after $2 \times 3 + 2$? Well, it's $2 \times 3 + 3$. Again we see that this number is composite because now 3 divides the first term, namely the product 2×3, and 3 divides the second term—the $+ 3$. So therefore, 3 divides the sum.

Notice that we've just rediscovered the two consecutive composite numbers 8 and 9. Now applying this method, we're ready to consider the more general case of finding a run of N consecutive composite numbers. We'll start with a number that I'll call K, defined to be the product of natural numbers given by $2 \times 3 \times 4 \times 5 \cdots$, all the natural numbers starting with 2 and ending with $(N + 1)$.

We observe that K is divisible by all the numbers from 2 out to $(N + 1)$, because it's a product of all of them. Notice that there are N numbers on this list of divisors, because we start with 2 and we end with $(N + 1)$, so that's a total of N numbers.

We now consider $K + 2$. That number is the product $2 \times 3 \times 4 \times 5 \cdots$, all the way out to $\times (N + 1)$, and then we add 2. The number $K + 2$ is divisible by 2 because that first product is—we have the 2 out in front, and that plus 2 has a factor of 2, and so we can see that that number is composite, because it's even.

The next number after $K + 2$ is $K + 3$, and that equals $2 \times 3 \times 4 \cdots$, out to $\times (N + 1) + 3$. We note that this number is also composite, because 3 is a factor of $K + 3$. Right? Three is a factor of that product, and then we add the 3—another 3—and so in fact we see that 3 divides this number evenly.

Similarly, we see that 4 is a factor of $K + 4$, and so it's composite. This pattern continues all the way up to the number $K + (N + 1)$, which is composite because $(N + 1)$ actually is a factor of the product 1, 2, 3, 4, ..., out to $(N + 1)$, and also that last term we're adding on: $+ (N + 1)$.

Therefore, we realize that each of the N consecutive numbers $K + 2$, $K + 3$, $K + 4$, ... , all the way out to finally $K + (N + 1)$, is composite. So we found a run of at least N consecutive composite numbers.

This argument is said to be effective in that it gives a method for actually finding the run of numbers—we could actually compute them—although it is not very practical for long runs because the product of all those numbers gets very large very quickly.

We remark that by creating a variation on Euclid's theme we were able to produce an entirely new result. Given that there are arbitrarily long gaps in which there are no primes, we are led to the feeling that there's no pattern within the prime numbers.

As we'll see in the upcoming lectures, the great Gauss made several important insights into the primes. Still, even with all these insights, even he was inspired to write the following:

> The problem of distinguishing prime numbers from composite numbers and of resolving the latter into their prime factors is known to be one of the most important and useful in arithmetic. It has engaged the industry and wisdom of ancient and modern geometers to such an extent that it would be superfluous to discuss the problem at length.

He continues:

> Further, the dignity of the science itself seems to require that every possible means be explored for the solution of a problem so elegant and so celebrated.

The study of the primes is, indeed, an elegant and celebrated field of study, one which we'll take up again in the next lecture.

Lecture Eight
Euler's Product Formula and Divisibility

Scope:

We will open this lecture by deriving what is arguably the most important formula involving prime numbers. In particular, we will connect an endless product involving the primes with the endless sum of the reciprocals of natural numbers: $1/1 + 1/2 + 1/3 + 1/4 + \cdots$. The derivation of this formula will involve our previous work on geometric series. This fundamental identity, first found by Leonhard Euler, will lead us to a "modern" proof that there are infinitely many primes. We will then see that while Euclid's ancient proof of the infinitude of primes is considered to be one of the most aesthetically appealing arguments in mathematics, Euler's analytic proof led naturally to the dawn of modern analytic number theory. While we address the true importance of Euler's formula and its generalizations in the lecture that follows this one, here we will explore how his formula allows us to analyze subtle questions involving divisibility of generic or randomly selected natural numbers. These questions are delicate and have a probabilistic feel to them. In particular, we will determine the likelihood that a natural number selected at random will have no repeated prime factors and, through our analysis, experience for ourselves the power of Euler's remarkable formula.

Outline

I. A formal formula of Euler's.

    A. Euler's amazing product formula.

        1. Leonhard Euler, an 18th-century Swiss mathematician, was one of the most prolific and important mathematicians in history.

        2. In 1737, he discovered a formula that gave birth to modern analytic number theory.

        3. This formula shows that a certain product involving prime numbers equals a certain sum of fractions.

        4. Euler's formula states that $(1/(1 - 1/2)) \times (1/(1 - 1/3)) \times (1/(1 - 1/5)) \times (1/(1 - 1/7)) \times (1/(1 - 1/11)) \times \cdots = 1 + 1/2 + 1/3 + 1/4 + 1/5 + 1/6 + \cdots$.

5. So Euler's product formula asserts that if we multiply $1/(1 - 1/p)$ for every prime number p, then the result will be the sum of the reciprocals of all the natural numbers.

6. We will now see why such a formula is believable by manipulating the product to equal the endless sum. We will treat both of these expressions as formal objects; that is, we will not wonder if these expressions represent actual numbers.

B. Returning to geometric infinite series.

1. We return to the formula we derived for geometric infinite series: For any number r satisfying $0 < r < 1$, we found that the endless sum $1 + r + r^2 + r^3 + r^4 + \cdots$ equals $1/(1 - r)$.

2. This is the key formula that will allow us to see why Euler's product formula is believable.

3. For any prime number p, we notice that $1/p$ is positive and less than 1 $(0 < 1/p < 1)$. The numbers we are multiplying together in Euler's formula are of the form $1/(1 - 1/p)$, so we now have a crucial insight: Numbers of that form are the sum of an infinite geometric series!

4. Specifically, we note that $1 + (1/p) + (1/p)^2 + (1/p)^3 + (1/p)^4 + \cdots = 1/(1 - 1/p)$.

C. A crash course in multiplication.

1. We now focus on the product in Euler's product formula and replace each term by its equivalent infinite geometric series.

2. We find: $(1/(1 - 1/2)) \times (1/(1 - 1/3)) \times (1/(1 - 1/5)) \times \cdots = (1 + (1/2) + (1/2)^2 + (1/2)^3 + \cdots) \times (1 + (1/3) + (1/3)^2 + (1/3)^3 + \cdots) \times (1 + (1/5) + (1/5)^2 + (1/5)^3 + \cdots) \times \cdots$.

3. To multiply these geometric series together, we pluck out one term from each sum and multiply those terms together. We then add up all the products we get from all the different ways of plucking one term from each geometric series.

4. For example, if we select 1 from each series, then that product of 1s equals 1. If we select 1/2 from the first series and 1s from the others, then that product equals 1/2. If we select 1 from the first series, 1/3 from the second series, and 1s from the rest, then that product equals 1/3.

5. Will we find 1/4? Yes—we select 1/4 from the first series and 1s from the rest. How about 1/5? Sure—select 1 from the first and second series, pluck out 1/5 from the third series, and 1s from the rest. How about 1/6? Yes, but this is a bit trickier: We pluck 1/2 from the first series, 1/3 from the second series, and 1s from the rest. That product yields 1/6.

6. Because we have all the powers of all the primes, then by the fundamental theorem of arithmetic we know that this process will produce all numbers of the form $1/n$ for all natural numbers n and that the reciprocal will only appear once in our sum.

7. For example, how would we find 1/300? We would pluck 1/4 from the first series, 1/3 from the second series, 1/25 from the third series, and 1s from the rest.

8. Thus we see that Euler's product formula, $(1/(1 - 1/2)) \times (1/(1 - 1/3)) \times (1/(1 - 1/5)) \times (1/(1 - 1/7)) \times (1/(1 - 1/11)) \times \cdots = 1 + 1/2 + 1/3 + 1/4 + 1/5 + 1/6 + \cdots$, makes sense. We caution that we are not claiming that either the endless product on the left or the endless sum on the right has actual numerical values.

D. Euler's general product formula.

1. In fact we can apply the same reasoning to verify a more general product formula due to Euler. Namely, for any number s, $(1/(1 - 1/2^s)) \times (1/(1 - 1/3^s)) \times (1/(1 - 1/5^s)) \times (1/(1 - 1/7^s)) \times (1/(1 - 1/11^s)) \times \cdots = 1 + 1/2^s + 1/3^s + 1/4^s + 1/5^s + 1/6^s + \cdots$.

2. The infinite series $1 + 1/2^s + 1/3^s + 1/4^s + 1/5^s + 1/6^s + \cdots$ is now known as the *zeta function* and is written as $\zeta(s) = 1 + 1/2^s + 1/3^s + 1/4^s + 1/5^s + 1/6^s + \cdots$.

II. A modern proof of the infinitude of primes.

A. The perplexing *harmonic series*.

1. Now that we have established Euler's product formula, we ask, what good is it?

2. The series $1 + 1/2 + 1/3 + 1/4 + 1/5 + 1/6 + \cdots$ is known as the *harmonic series* and is a very important object from calculus. We notice that each term in the sum is

smaller than its predecessor, and the terms are getting arbitrarily small (so they are approaching 0).

    3. This sum is important because despite the fact that the terms are shrinking to 0, the sum itself is *infinite*. When an infinite series does not sum to a number, we say that the series *diverges*.

    4. The harmonic series diverges, which seems counterintuitive, so we ask our standard question, why?

B. A divergent sum.

    1. A sneaky grouping of the terms in the harmonic series offers us some intuition about why the series sums to infinity (that is, diverges).

    2. We first collect the numbers in groups of sizes that equal ever-higher powers of 2. That is, we group the series $1 + 1/2 + 1/3 + 1/4 + 1/5 + 1/6 + 1/7 + 1/8 + \cdots$ as: $1 + (1/2) + (1/3 + 1/4) + (1/5 + 1/6 + 1/7 + 1/8) + \cdots$.

    3. We notice that this series is term-by-term larger than the series $1 + (1/2) + (1/4 + 1/4) + (1/8 + 1/8 + 1/8 + 1/8) + \cdots$.

    4. This "smaller" series equals: $1 + 1/2 + 1/2 + 1/2 + \cdots$. Adding infinitely many halves together yields an infinite answer. Thus the "larger" harmonic series must be infinite as well.

C. Another argument showing the infinitude of the primes.

    1. We now see why our argument to verify Euler's formula was "formal": The two expressions are indeed *not* numbers but *are* equal as mathematical expressions.

    2. Using our new insight that the harmonic series diverges, we can apply Euler's product formula to give a modern proof that there are infinitely many prime numbers.

    3. This proof is by contradiction: We will assume that there are only finitely many prime numbers and argue that this assumption leads to a contradiction—a logical fallacy.

    4. If there were only finitely many prime numbers, then the product in Euler's formula, $(1/(1 - 1/2)) \times (1/(1 - 1/3)) \times (1/(1 - 1/5)) \times \cdots$, would have a last term. Thus we would have a product of a finite number of rational numbers (fractions), and that product would be another rational number.

5. However, by Euler's product formula we know that this product equals the harmonic series, which we just showed equals infinity. Therefore we are forced to conclude that there is a rational number (a ratio of two natural numbers) that equals infinity. This is impossible, and thus we have reached a contradiction.

6. This fallacy implies that our original assumption must have been false; hence, there must be infinitely many primes.

7. Notice how different this argument is from Euclid's original proof that there are infinitely many primes.

III. The likelihood that a number is "square free."

  A. The likelihood of stumbling on an odd number.

    1. Euler's general product formula has important implications into the study of primes.

    2. To illustrate one implication, we consider a very simple question: What is the probability that a randomly selected natural number is odd?

    3. We consider the *opposite* question: What is the probability that this number is even (i.e., a multiple of 2)? Since every other number is a multiple of 2, the probability of the number being even equals 1/2. To find the opposite probability, we compute $1 - 1/2 = 1/2$. So we conclude that the probability that the random number is odd is 1/2.

    4. This elementary analysis can be extended to answer a much more interesting and subtle number question.

  B. No repeated prime factors: What are the chances?

    1. Moving beyond the prime numbers—those numbers that have just one number appearing in their prime factorization—we consider those numbers whose prime factorizations consist of no repeated primes. That is, those numbers for which the primes appearing in their prime factorizations appear only *once*.

    2. These numbers are called "square free." For example, 6 is square free because no repeated primes appear in its prime factorization, 2×3. However 20 is *not* square free because it does have a repeated factor in its prime factorization, $2 \times 2 \times 5$. Similarly, 54 is not square free because $54 = 2 \times 3 \times 3 \times 3$.

3. If we were to pick a natural number at random, what is the probability that it is square free, that is, *not* a multiple of 2^2, 3^2, 5^2, 7^2, 11^2, and so forth, for each prime number?

4. Applying our thinking for the question of *not* being a multiple of 2, we see that the likelihood of *not* being a multiple of 2^2 equals $1 - 1/2^2$; the likelihood of *not* being a multiple of 3^2 equals $1 - 1/3^2$; and in general for any prime number p, the likelihood that a random number is *not* a multiple of p^2 equals $1 - 1/p^2$.

5. We will string all these probabilities together with multiplication as we would if we were flipping a coin again and again—that is, we will view the likelihood with respect to each prime as independent.

6. So the probability of having no prime appearing twice is: $(1 - 1/2^2) \times (1 - 1/3^2) \times (1 - 1/5^2) \times (1 - 1/7^2) \times \cdots$, which is the reciprocal of the product from Euler's general product formula. Therefore this product equals $1/(1 + 1/2^2 + 1/3^2 + 1/4^2 + 1/5^2 + 1/6^2 + 1/7^2 + \cdots)$, which can be written as $1/\zeta(2)$.

C. A very special sum.

1. Unlike the harmonic series, the terms from the infinite series $\zeta(2)$ do shrink to 0 fast enough so that the entire series does converge to a number.

2. This infinite series can be computed through some advanced mathematical ideas from calculus. The answer is $\zeta(2) = \pi^2/6$.

3. So the probability that a random natural number is square free is $1/\zeta(2) = 6/\pi^2 = 0.6079\ldots$, or about 61%.

4. Euler's product formula appears in unexpected places throughout mathematics and science.

Questions to Consider:

1. What is the probability that a natural number chosen at random is *not* a multiple of 7?

2. We have seen that the harmonic series $1 + 1/2 + 1/3 + 1/4 + 1/5 + \cdots$ diverges to infinity. How many consecutive terms do you need to add to get a partial sum that exceeds 2? How about 3?

Lecture Eight—Transcript
Euler's Product Formula and Divisibility

This promises to be a very exciting lecture because we'll begin by deriving what is arguably the most important formula involving prime numbers.

In particular, we'll connect an endless product involving the primes with the endless sum of the reciprocals of natural numbers: $1/1 + 1/2 + 1/3 + 1/4 + 1/5$, and so forth. In fact, our derivation of this formula will incorporate our previous work on geometric series.

This fundamental identity, first found by the great 18th-century Swiss mathematician Leonhard Euler, will lead us to a "modern" proof that there are infinitely many primes. We'll then see that while Euclid's ancient proof of the infinitude of primes is considered to be one of the most aesthetically appealing arguments in mathematics, Euler's analytic proof gave way to the dawn of modern analytic number theory.

While we'll address the true importance of Euler's formula and its generalizations in the lecture that follows this one, here we'll explore how Euler's formula allows us to analyze subtle questions involving divisibility of generic, or randomly selected, natural numbers.

These questions are indeed delicate and have a probabilistic feel to them. In particular, we'll determine the likelihood that a natural number selected at random will have no repeated prime factors and through our analysis experience for ourselves the power of Euler's remarkable formula.

Leonhard Euler was an 18th-century Swiss mathematician and is considered to be one of the most prolific and important mathematicians in human history. Here we introduce the celebrated formula that he discovered in 1737, which gave birth to modern analytic number theory.

This formula shows that a certain product involving prime numbers equals a certain sum of fractions. It's a very strange-looking formula upon first inspection, but we'll parse it for ourselves and see exactly what it means.

Euler's formula states that if we take the reciprocal of $1 - 1/2$, times the reciprocal of $1 - 1/3$, times the reciprocal of $1 - 1/5$, times the reciprocal of $1 - 1/7$, times the reciprocal of $1 - 1/11$, and so forth for each prime, then that product will equal the sum $1 + 1/2 + 1/3 + 1/4 + 1/5 + 1/6$, and so on.

So Euler's product formula asserts that if we multiply the reciprocals of $(1 - 1/p)$ for every prime number p then the result will be the sum of the reciprocals of all the natural numbers.

We'll now see why such a formula is believable, by manipulating the product, and show that it actually can be manipulated to equal the endless sum. We'll treat both of these expressions as formal objects. In other words, we'll use algebra to simplify the product, but we won't wonder, at this point, if these expressions represent actual numbers.

To begin with, we'll return to our explorations into geometric series, in particular, the formula we derived for geometric infinite series. Let me remind you that if we take a ratio r that's a positive number but less than 1, then we found that the endless sum $1 + r + r^2 + r^3 + r^4 + \cdots$, and so on, forever, actually can be summed, and we have this wonderful closed formula $1/(1 - r)$. This is the key formula that will allow us to see why Euler's product formula is indeed believable.

For any prime number p, let's notice that its reciprocal, $1/p$, is first of all positive and less than 1. The numbers we're multiplying together in Euler's formula are of the form $1/(1 - 1/p)$, so we now make a crucial observation: Numbers of that form are the sum of an infinite geometric series, and that's the secret.

Specifically, we note that the reciprocal of $1 - 1/p$ actually equals— by our previous work—$1 + (1/p) + (1/p)^2 + (1/p)^3 + (1/p)^4 + \cdots$, and so on.

Let's now embark on a crash course in multiplication. We focus on the product in Euler's product formula, and either in a stroke of genius or madness we'll replace each term by its equivalent infinite geometric series. So this is going to be a real mouthful here, but let's think about it together.

What we're going to do is we're going to see that if we take the reciprocal of $(1 - 1/2)$ and multiply it by the reciprocal of $(1 - 1/3)$, times the reciprocal of $(1 - 1/5)$, and we do this for every prime, then we could write that as the following product: The product that first consists of the sum $1 + (1/2) + (1/2)^2 + (1/2)^3 + (1/2)^4 + \cdots$, and so on, all multiplied by the next term, which would be $1 + (1/3) + (1/3)^2 + (1/3)^3 + (1/3)^4 + \cdots$, and so on, all multiplied by $1 + (1/5) + (1/5)^2 + (1/5)^3 + \cdots$, and so forth, down the line for all the primes.

To multiply these infinite geometric series together, what we do is we pluck out one term from each of the sums and multiply those terms together. We then add up all the products we get from all the different ways of plucking one term out of each of these geometric series, and that's the product. Let's take a look at some examples.

If we select 1 from each series and multiply all those 1s together, we just see $1 \times 1 \times 1$, and so we get 1. So the very first term we get when we multiply just the 1s together is the 1.

What if we select 1/2 from the first series and 1s from all the others? Then when I multiply them together, I'll see 1/2 times an infinite run of 1s that we multiply together to just get 1/2.

What if we select 1 from the first series, and then 1/3 from the second series, and then 1s all from the rest? Well, then we see that product equals 1/3.

Will we find 1/4 in our sum? Yes. What we do is we select 1/4 from the first series—which, by the way, notice is $(1/2)^2$—and we select 1s from all the rest. When I take the product, I see 1/4. What about 1/5? Sure. No problem. Select 1 from the first series, select 1 from the second series, and then pick out 1/5 from the third series, and then 1s all the rest of the way, and we see 1/5.

How about the number 1/6? The answer is, yes, we'll see it, but it's a little bit trickier. Here's how we get 1/6: We pluck out 1/2 from the first series, and then 1/3 from the second series, and then 1s from all the rest. That product yields 1/6.

Because we have the reciprocals of all the powers of all the primes, then by the fundamental theorem of arithmetic that we discussed in the previous lecture, we know that this process will produce all numbers of the form $1/n$ for all natural numbers n, and by the uniqueness of the factorization, the reciprocal we see will only appear once in our sum.

Let me try to illustrate this with one last example. How would we find the term 1/300? First we factor 300. So, $300 = 4 \times 3 \times 25$. So 1/300 is $1/4 \times 1/3 \times 1/25$. Therefore we would pluck 1/4 out from the first series, which is $(1/2)^2$, we'd pluck the term 1/3 out from the second series, and from the third series we'd pluck out $(1/5)^2$—which is 1/25—and then 1s all the rest of the way. Notice that that product is indeed 1/300. So we would add that in with the mix.

Thus we see that Euler's product formula, the reciprocal of $(1 - 1/2)$ times the reciprocal of $(1 - 1/3)$, and so forth for all the primes—if you multiply all those together, that will equal $1 + 1/2 + 1/3 + 1/4 + 1/5 + 1/6$, and even all the way out to $+ 1/300$, and plus, and so forth, forever. And now this makes sense, kind of. It takes a while to grasp it, though.

As complicated as that is, I want to caution that we are not claiming that either the product on the left or the endless sum on the right represents an actual numerical value. We'll return to this important point in just a moment.

Applying the same reasoning as we just found, we could actually verify a slightly more general product formula that's also due to Euler. Namely, for any number—let me just call it s—he proves that the reciprocal of $(1 - 1/2^s)$ times the reciprocal of $(1 - 1/3^s)$ times the reciprocal of $(1 - 1/5^s)$ times the reciprocal of $(1 - 1/7^s)$, and so forth—take the product of all these primes, but now raised to the s power—then if we multiply all that out using the exact same idea, we see a sum, but now the sum is $1 + 1/2^s + 1/3^s + 1/4^s + 1/5^s + 1/6^s$, all the way out to $1/300^s$, plus, and so on, all the way out.

The infinite series that we just saw here, $1 + 1/2^s + 1/3^s + 1/4^s + 1/5^s + 1/6^s + \cdots$, and so forth, is now known as the *zeta function* and is written using the Greek letter ζ [zeta]. We say $\zeta(s)$ equals this particular infinitely long series of numbers.

Now that we've established Euler's product formula—and by the way, that does take a while to absorb because there's lots of products that we're putting together to find the sum—we might very well ask a totally fair question, what in the world is it good for? That's a wonderful, fair question.

First of all, the series that we saw: $1 + 1/2 + 1/3 + 1/4 + 1/5 + 1/6 + \cdots$, and so forth, forever, is actually known as the *harmonic series*. It is a very important object from calculus.

Let's observe that each term in the sum is smaller than its predecessor, and the terms are getting arbitrarily small. In other words, the terms individually are actually shrinking down to 0. This sum is important because despite the fact that the individual terms themselves are shrinking towards 0, the entire sum, taken in its entirety, is *infinite*. When an infinite series does not sum to a number we say that the series *diverges*, or in this case we'd say the series diverges to infinity.

The harmonic series diverges, which seems counterintuitive. So we have to ask the standard question, why? Why does this infinite list of ever-shrinking numbers actually diverge—meaning that it doesn't actually sum to a number.

I'm going to show you an argument that verifies this, but what follows is a little bit tricky, so let's not get bogged down in the details. If you're having fun with the details, great, and if not, let's just let it go. But here's a sneaky grouping of the terms in the harmonic series that offers us some intuition why the series actually sums to infinity. In other words, diverges.

We first collect the numbers and collect them into special groupings. So, here's the grouping of the infinite series that I want us to consider. We first start off with $1 + 1/2 + 1/3 + 1/4 + \cdots$, and so on, and now I want to group them as follows: I want to have 1 by itself, and then I want to group 1/2 by itself. Now I'm going to pick the next two terms, which is $1/3 + 1/4$, and I want to group them together.

Then I'm going to group the next four terms together, so $1/5 + 1/6 + 1/7 + 1/8$, and so on. Then I'm going to group the next eight terms together, which would be 1/9 and so forth. What you'll see here is I'm grouping groups of powers of 2 together. So first I have 2, then I have 4, then I'll have 8, and so forth.

Now, let's take a look at the first grouping, the $1/3 + 1/4$. Notice that the fraction 1/3 is actually a larger quantity than the fraction 1/4. So, 1/3 is bigger than 1/4. And if we replace the 1/3 in the sum by a 1/4, we would create a smaller sum. I'd replace the bigger number—1/3—by the slightly smaller number—1/4.

So, now in that first grouping, I would see a $1/4 + 1/4$. By the way, $1/4 + 1/4$ is 2/4, which equals 1/2. So I see now that the new sum, which is smaller, starts with $1 + 1/2$, and now that first grouping of terms really shrinks down to just 1/2.

Now let's look at the next grouping, that's the grouping of the four terms: $1/5 + 1/6 + 1/7 + 1/8$. Notice that the fractions $1/5$, $1/6$, $1/7$, are all individually larger than the fraction $1/8$; $1/8$ is the smallest. Thus, if we replace those three fractions—$1/5$, $1/6$, and $1/7$—each by another copy of $1/8$, we would again take our sum and actually make it even smaller still.

In this case, that grouping would now be replaced by $1/8 + 1/8 + 1/8 + 1/8$, which equals $4/8$, or again, $1/2$.

So the next grouping of numbers actually is now $1/2$. Interesting, we keep seeing these $1/2$s occurring. In fact, this is going to be a recurring theme.

If we continue this replacement process down the line, we would see that the harmonic series—the original series—is larger than the series that we're now creating, which is $1 + (1/2) + (1/4 + 1/4) + (1/8 + 1/8 + 1/8 + 1/8) + (1/16 + 1/16 + 1/16$, eight times$)$, and so forth. This new "smaller" series we just created actually equals $1 + 1/2 + 1/2 + 1/2 + 1/2 + \cdots$, forever.

Since there's no end to these groupings, each of which gives $1/2$, we see that adding infinitely many halves together yields an infinite answer. If you take $1/2$ and add it to itself forever, we're going to get something that's infinite. Thus the "larger" harmonic series must be infinite as well. Hence the harmonic series diverges to infinity.

That's an awful lot to grasp, but I invite you to think about it if it's something that interests you, and if not, then let's just see that we've taken the harmonic series and shown that we could produce something that seems smaller yet is infinite; therefore, the harmonic series must itself be infinite as well.

Now, see why I cautioned early in our argument as we verified Euler's formula that it was a "formal" formula. The two expressions are indeed *not* numbers but *are* equal as formal mathematical expressions. So, this is a very unusual point. But using our new insight—namely the insight that the harmonic series diverges to infinity—we can apply Euler's product formula to provide a modern proof that there are infinitely many prime numbers. So, even if you weren't so hot on that derivation about the divergence of the harmonic series, I want you to come back here, because this is really neat. This is really, really neat.

We're going to now prove that there are infinitely many primes, but using this formula. This proof is by contradiction, which means that we'll assume that there are only finitely many prime numbers, and we'll argue that this assumption leads to a contradiction—in other words, a logical fallacy.

If there were only finitely many prime numbers, then the product in Euler's formula—remember that product on the left: the reciprocal of $(1 - 1/2)$ times the reciprocal of $(1 - 1/3)$ times the reciprocal of $(1 - 1/5)$, and so on—well, there would be a last factor in that product. Thus, we'd have a product of a finite number of rational numbers—a finite number of fractions. That product would be another fraction.

However, by Euler's product formula we know that this product equals a harmonic series, which we just demonstrated equals infinity. Therefore, we are forced to conclude that there is a rational number—a ratio of two natural numbers—that equals infinity. This is obviously impossible, and thus we've reached a contradiction.

This fallacy implies that our original assumption must have been false; hence there must be, in fact, infinitely many primes. Notice how different this argument is from Euclid's original proof that there are infinitely many primes. Absolutely wonderful. This formula actually gives us some information about the primes, albeit information that we already knew from ancient Greek mathematics.

Euler's general product formula has important implications in the study of primes, and I want to illustrate one of these implications right now. But to begin with, I want to consider a very, very simple question and then slowly build up.

Suppose we randomly pick a natural number. What is the probability that this number is odd? To answer this question, we consider the *opposite* question, what is the probability that this number is even? A number is even precisely when it's a multiple of 2. We've already seen this.

Since every other number is a multiple of 2, the probability that the number is even equals 1/2. To find the probability of the opposite scenario—the opposite probability—we compute $1 - 1/2$, which still equals 1/2. So we conclude that the probability that the random number is odd equals 1/2, which maybe is not so startling. Maybe that intuitively made sense to you. But I wanted us to think about the reasoning in this way because we're going to see this reasoning actually allows us to do something else. In fact, this elementary analysis can be extended to answer a much more interesting and subtle number theory question.

Moving beyond the prime numbers—those numbers that have just one number appearing in their prime factorization—we now consider those numbers whose prime factorizations consist of no repeated primes. That is, those numbers for which the primes appearing in its prime factorizations appear only *once*.

These numbers are called "square free," and in some sense produce the next level of generality after the primes—namely, the numbers that just have the primes, multiples appearing just once. For example, 6 is square free because no repeated primes appear in its factorization. We just see 2×3—both primes appear once.

However 20 is *not* square free because it does have a repeated factor in its prime factorization. Notice that 20 equals $2 \times 2 \times 5$, and we see a repeated factor. Similarly, a number like 54 is not square free because if we factor 54, we'd see $2 \times 3 \times 3 \times 3$. So we see the repeated factor of 3, and therefore it's not square free.

If we pick a number at random, a natural number at random, what is the probability that it's square free? In other words, what's the probability that it's *not* a multiple of 2^2, not a multiple of 3^2, not a multiple of 5^2, not a multiple of 7^2, not a multiple of 11^2, and so forth down the line, for each and every prime number?

Applying our method for answering the easy question of a number *not* being a multiple of 2, we see that the likelihood of *not* being a multiple of 2^2 equals $1 - 1/2^2$. The likelihood of *not* being a multiple of 3^2 equals $1 - 1/3^2$. In general, for any prime number p, the likelihood that a random natural number is *not* a multiple of p^2 equals $1 - 1/p^2$.

We'll find our answer by multiplying all these individual probabilities together. Let's pause for a moment to see why multiplying these probabilities makes sense. Suppose we flip a fair coin. What's the probability of seeing tails? Well, it's 1/2, because there are two possibilities—heads, tails—and they're equally likely, so 1 out of 2. Well, what if we flip the coin twice? The coin has no memory, it doesn't know how it landed one way and it doesn't know how it's going to land the next way, so there's no memory. It doesn't record what it's done. So, the probability that we see 2 tails in a row would actually be 1/2 multiplied by 1/2, which would be 1/4. In fact, the probability of flipping a coin twice and seeing a tail followed by a tail is, in fact, 1 out of 4.

Similarly, returning to our square-free probabilities, we string all our probabilities together with multiplication as if we were flipping a coin again and again and again. In other words, we'll view the likelihood with respect to each prime as independent. So the probability of having no prime appearing twice is the product $(1 - 1/2^2) \times (1 - 1/3^2) \times (1 - 1/5^2) \times (1 - 1/7^2)$, and so forth. We do this for each prime.

Notice that this product is the reciprocal of the product from Euler's general product formula, with $s = 2$. Therefore, this product equals—by the product formula—the reciprocal of the sum of the squares of all the natural numbers. In particular, the product equals the reciprocal of $1 + 1/2^2 + 1/3^2 + 1/4^2 + 1/5^2 + 1/6^2 + 1/7^2 + \cdots$, and so forth—the reciprocal of the square of every natural number.

This number can actually be written as $1/\zeta(2)$, because remember, we use the ζ symbol to represent this sum, and since we're letting the exponent be 2, we see this as $1/\zeta(2)$.

Unlike the harmonic series that we saw earlier, here the terms from the infinite series $\zeta(2)$ do shrink to 0 fast enough so that the entire infinite series does converge upon a number, just like we saw with the geometric infinite series.

This infinite series can be computed precisely using some advanced mathematical ideas from calculus. But let me just tell you that the answer is $\zeta(2)$—that infinite series—equals π^2 divided by 6. Remember that π, which we saw from geometry way back in our school days, is equal to around 3.1415…, and so forth. So we're looking at that number squared, divided by 6.

Returning to our question, we see that the probability that a random natural number is square free is actually the reciprocal of $\zeta(2)$, which is the reciprocal of $\pi^2/6$, namely, $6/\pi^2$. We could use a calculator to see what this number looks like in decimal form, and we see it's 0.6079…, and it goes on, or approximately a probability of 61%.

This is amazing, because what we're saying is that if you pick a number at random it's more likely than not—in fact, approaching a two-thirds chance—that it's going to be square free. So the square-free numbers seem to be more popular within the natural numbers than the numbers that are not square free. So, a new insight.

We can actually try an example to see if this number is at least reasonable, or believable. We could count how many numbers less than or equal to 100 are square free. So, let's just consider picking a random number from 1 to 100. Forget about picking a random number from 1 to forever. Now you can actually use a sieve-type method that we saw in the previous lecture, and we could actually see how many of these square-free numbers there are.

In fact, if you try the sieve method on your own, you need only cross off the multiples of 4, 9, 25, and 49, and I'll let you think about that. But if we did this—and I did this already—we'd discover that there are in fact 61 square-free numbers from the first 100 natural numbers. So the probability of picking a square-free number out of the first 100 natural numbers is 61 out of 100, or 0.61, or 61%, which is amazingly close to the actual value we computed when we pick among any natural number of 60.79%, which we just found.

So, it is more likely than not that a natural number picked at random will have no repeated factors. Absolutely amazing that we could find that out. We've just made a new insight into the infinite collection of natural numbers. Even though we can't say things about all of the numbers taken at large—we don't know all those big prime numbers—we can say something precise about the totality if we pick a number out at random.

Euler's product formula not only provides important insights into the prime numbers but actually appears in unexpected places throughout mathematics and science.

In the next lecture, we'll see how Euler's formula leads to one of the most famous long-standing open questions in all of mathematics. Euler himself wondered about a pattern within the primes. His words today still ring true. He wrote, "Mathematicians have tried in vain to this day to discover some order in the sequence of prime numbers, and we have reason to believe that it is a mystery into which the mind will never penetrate."

So we see that the study of primes is indeed an enormous challenge, and in the next lecture we'll see some open questions that still tantalize practitioners of number theory—to this very day and into the future.

Lecture Nine
The Prime Number Theorem and Riemann

Scope:

In this lecture we close our explorations into analytic number theory by studying its crown jewel, the prime number theorem, which answers a question of great interest to mathematicians: Can we estimate how many primes there are up to a certain size? To improve the estimates within this important result we will return to Euler's general product formula and come upon one of the most important unsolved problems in all of mathematics—the Riemann Hypothesis. The truth of the Riemann Hypothesis would immediately imply a long list of deep results. While the issues surrounding the Riemann Hypothesis at first appear to be totally divorced from our "real world," we will mention some tantalizing new connections with physics through what are known as *random matrices*, which might hold the key to unlocking the mysteries of this long-standing open question about the atoms of the natural numbers—the primes. While a complex proof of the prime number theorem was found in 1896, many number theorists wondered if there was an "elementary" argument. Finally, in 1948 such a proof was found by two great mathematicians: Paul Erdös and Atle Selberg. This mathematical milestone led to one of the most famous disputes in number theory history. We will close this chapter of our course by describing some modern advances in the study of primes, some of which will allow us to return to the concept of arithmetic progressions. Finally we will ponder some famous questions involving prime numbers that remain mysteries to this very day.

Outline

I.   The prime number theorem.

    A. How many primes are there?

        1. We close our study of the primes by returning to Euclid's theorem stating that there are infinitely many primes.

        2. Can we make this result more precise?

B. A remarkable estimate.

    1. We write $\pi(n)$ to denote the number of primes less than or equal to n.

    2. So, for example, $\pi(25) = 9$ because there are nine primes that are less than or equal to 25: 2, 3, 5, 7, 11, 13, 17, 19, and 23.

    3. Euclid's theorem asserts that $\pi(n)$ approaches infinity as n gets larger and larger.

    4. Is there a formula for $\pi(n)$? This remains an open question seemingly impossible to answer.

C. An "elementary" proof.

    1. By the late 18th century, French mathematician Adrien-Marie Legendre and the great Gauss noticed that the number of primes less than or equal to n, $\pi(n)$, seemed to be connected with the "natural logarithm" function, $\ln(n)$.

    2. The natural logarithm of a natural number n can be roughly viewed as approximately the number of digits n contains.

    3. Thus we see that $\ln(n)$ is a very slow-growing function.

    4. The conjecture was as n gets larger and larger, $\pi(n)$ gets closer and closer to $n/\ln(n)$.

    5. The great Russian mathematician Pafnuty Chebyshev proved in 1850 that *if* the quantity $\pi(n)/(n/\ln(n))$ approaches a number as n gets larger and larger, then that number must equal 1.

    6. In 1859 the great German mathematician Bernhard Riemann introduced a number of revolutionary ideas in his memoir *On the Number of Primes Less Than a Given Magnitude*. Among other things, he showed how this issue is connected with complex numbers (numbers involving the imaginary number $i = \sqrt{(-1)}$) and the zeta function, $\zeta(s)$, that we saw in Euler's general product formula. Today the zeta function is known as the *Riemann zeta function*.

    7. In 1896 French mathematician Jacques Salomon Hadamard and Belgian mathematician Charles de la Vallée-Poussin independently produced a proof of the prime number theorem: As n gets larger and larger, the number of primes up to n approaches $n/\ln(n)$. More precisely, as n gets larger and larger, the ratio $\pi(n)/(n/\ln(n))$ approaches 1.

II. The Riemann Hypothesis.

A. The "error" in the prime number theorem.

1. The prime number theorem implies that as n gets larger and larger, $\pi(n)$ gets closer and closer to $n/\ln(n)$.

2. However for any particular n, $\pi(n)$ is not equal to $n/\ln(n)$; there is an error given by the difference between these two numbers.

3. How close are these two numbers in actuality for larger and larger values of n?

B. Bernhard Riemann and his famous hypothesis.

1. In Riemann's famous work, he found a profound connection between $\pi(n)$ and the zeta function: $\zeta(s) = 1 + 1/2^s + 1/3^s + 1/4^s + 1/5^s + 1/6^s + 1/7^s + \cdots$.

2. Riemann's insight was to extend the series $\zeta(s)$ to allow s to be a complex number, that is, $s = x + iy$, for real numbers x and y, and $i = \sqrt{(-1)}$.

3. Riemann studied the complex numbers s that were solutions to the following strange-looking equation: $\zeta(s) = 0$. Or, equivalently but even stranger: $1 + 1/2^s + 1/3^s + 1/4^s + 1/5^s + 1/6^s + 1/7^s + \cdots = 0$.

4. He proved that all the complex solutions $s = x + iy$ satisfy the condition that $0 \leq x \leq 1$ and showed that if these bounds could be tightened, then the error term in the prime number theorem could be reduced.

5. A key step in Hadamard's and de la Vallée-Poussin's proof of the prime number theorem was showing that all solutions satisfied $0 < x < 1$.

6. Riemann conjectured that *all* the solutions satisfied $x = 1/2$. This conjecture is now known as the *Riemann Hypothesis*.

C. What if it's true?

1. If the Riemann Hypothesis were true, then we would have a much smaller error term in the prime number theorem. In addition, we would learn an enormous amount about the prime numbers because there are

literally hundreds of theorems that begin "Assuming the truth of the Riemann Hypothesis … ."

2. In 1900 the great German mathematician David Hilbert included the Riemann Hypothesis on his list of the 23 most important unsolved questions in mathematics. One hundred years later, the Clay Mathematics Institute in Cambridge, MA, listed it as one of its "Millennium Problems"—a correct and complete answer would result in a prize of $1 million.

3. It remains one of the most important open questions in all of mathematics.

D. Current state of knowledge.

1. There is a long list of partial results surrounding the Riemann Hypothesis.

2. From 2001 through 2005, a number theory program called "ZetaGrid" verified that the first 100 billion solutions to the equation had $x = 1/2$. Of course, this overwhelming data is not a general proof.

3. A new direction toward a possible proof of the Riemann Hypothesis is to study objects called *random matrices*. These are mathematical objects that were originally applied to better understand the quantum behavior of larger atoms in physics.

III. The drama behind an "elementary" proof of the prime number theorem.

A. A search for an "elementary" proof.

1. The original proof of the prime number theorem was extremely clever and subtle (especially in demonstrating that $x < 1$ for all the "zeros" of the Riemann zeta function).

2. In 1921 the great British analytic number theorist G. H. Hardy wondered if an "elementary" proof of the prime number theorem could be found—that is, a proof that involved only very simple properties of functions but in an extraordinarily ingenious manner.

3. Many believed that such an "elementary" proof might not be possible.

B. Paul Erdös, Atle Selberg, and their contributions.

   1. In 1948 Paul Erdös announced that he and Atle Selberg had found a truly "elementary" proof that involved only basic properties of the logarithm.

   2. This announcement stunned the mathematical world.

   3. Paul Erdös was a Hungarian mathematician who published well over 1500 mathematical articles with over 500 coauthors from around the world. He was also a nomad, having no true academic affiliation for most of his life.

   4. His prolific and foundational work inspired and fed new branches of mathematics including *graph theory*, *combinatorial number theory*, and *elementary number theory*.

   5. Atle Selberg was a Norwegian mathematician who spent much of his career at the Institute of Advanced Study at Princeton.

   6. He produced profound results in advanced areas of number theory including automorphic forms, and he introduced several new areas of study including the "Selberg sieve" and the "Selberg trace formula."

C. Their important proof.

   1. Their "elementary" proof of the prime number theorem was so sensational that in 1950 Selberg was awarded the Fields Medal (the mathematical equivalent of the Nobel Prize).

   2. In 1952 Erdös received the Cole Prize (one of the most prestigious prizes in mathematics).

D. The controversy that ensued.

   1. A serious controversy arose over who should receive credit for what part of the proof.

   2. In March 1948, Selberg discovered an important formula involving primes but did not publish it.

   3. Several months later, Selberg shared with Hungarian mathematician Paul Turan an inequality he discovered (now known as the *fundamental formula*). Without Selberg's objection, Turan gave a lecture outlining Selberg's recent work.

4. **Erdös**, who was in the audience, quickly exclaimed, I think you can also derive lim $p_{n+1}/p_n = 1$ (as n approaches infinity) from this inequality.

5. Within a few hours Erdös produced an ingenious proof of his extremely important assertion. A day later, when Erdös shared this news with Selberg, Selberg responded with, You must have made a mistake.

6. A few days later, Selberg, using his formula involving logarithms and primes together with Erdös's important result, was able to devise an "elementary" proof of the prime number theorem. The key ingredient, however, was Erdös's theorem.

7. Erdös suggested that the two collaborate and write a joint paper, but Selberg suggested that each write their own papers based on their own work. Erdös found this objectionable.

8. To make matters worse, news started to spread about this amazing breakthrough, but the rumor attributed the result to Erdös. In the fall of 1948, someone greeted Selberg with, Have you heard the exciting news of what Erdös has proven? This did not help the situation.

9. The two did publish their papers separately, including mention of each other's work. However, the two did not speak to each other again for 45 years.

IV. Further advances on the distribution of the primes.

    A. Dirichlet's theorem on primes in arithmetic progressions.

        1. We recall the idea of an arithmetic progression: We start with a number and continue to add a fixed amount. For example, start with 1 and repeatedly add 17 to obtain 1, 18, 35, 52, and so forth.

        2. Gauss wondered if such arithmetic progressions always contained prime numbers. In this example we have 1, 18, 35, 52, 69, 86, 103, … , and 103 is the first prime number we found.

        3. In 1837, German mathematician Johann Dirichlet proved that given any natural numbers A and B not sharing any prime factors, the arithmetic progression A, A + B, A + 2B, A + 3B, A + 4B, and so on, will always contain infinitely many prime numbers. This extends Euclid's theorem on the infinitude of primes.

B. Arithmetic progressions of primes.
   1. Suppose we list all the prime numbers in order: 2, 3, 5, 7, 11, 13, 17, 19, ...
   2. Within this list, do we see any pieces that form an arithmetic progression? For example, 3, 5, 7 forms an arithmetic progression (we repeatedly add 2). What is the *longest* arithmetic progression found in the list of primes?
   3. In 2004 some groundbreaking work was done in this direction by Ben Green and Terence Tao. They proved the astounding result that there are arbitrarily long arithmetic progressions within the list of prime numbers.

V. Famous open questions involving prime numbers.

   A. Twin primes. Are there infinitely many pairs of primes that differ by 2? For example, (3, 5), (5, 7), (11, 13). These pairs are known as *twin primes*. Notice that (13, 17) is not a pair of twin primes because their difference is not 2: $17 - 13 = 4$. The Twin Prime Conjecture states that there are infinitely many twin primes. But a proof continues to elude us.

   B. Goldbach's conjecture.
   1. We notice that $6 = 3 + 3$ (the sum of two primes); $8 = 3 + 5$; $10 = 5 + 5$; $12 = 5 + 7$; $14 = 7 + 7$; and $16 = 5 + 11$. Here we see that each of these even numbers can be expressed as the sum of two primes.
   2. The Goldbach conjecture states that every even number greater than 2 can be written as the sum of two prime numbers.
   3. The conjecture was first made by Christian Goldbach in a letter to Euler dated June 7, 1742.
   4. Even though the conjecture has been shown to hold for all even numbers up to 3×10^{17}, a proof that it holds for *all* even numbers has yet to be found.

   C. Skewes number.
   1. We have seen that $\pi(n)$ approaches the function $n/(\ln(n))$. However, there is another function that behaves like $n/(\ln(n))$ and also approximates $\pi(n)$. This is called the *logarithmic integral* and is written $\text{Li}(n)$. It has been shown that $\pi(n) - \text{Li}(n)$ approaches 0 as n gets larger and larger.

2. In 1914 British mathematician John Littlewood proved that the quantity $\pi(n) - \text{Li}(n)$ changes from positive to negative infinitely many times.
3. However, for all values up to around 10^{22}, the quantity has been negative (that is, $\text{Li}(n)$ has been larger than $\pi(n)$). Given Littlewood's result, we know that $\text{Li}(n)$ must at some point be smaller than $\pi(n)$. But when?
4. South African mathematician Samuel Skewes in 1933 proved that, assuming the truth of the Riemann Hypothesis, $\text{Li}(n)$ must be smaller than $\pi(n)$ for some n less than $10^{10^{10^{34}}}$. Now known as *Skewes number*, it was described by G. H. Hardy as, "the largest number which has ever served any definite purpose in mathematics."

Questions to Consider:

1   Can there be an arithmetic progression of three consecutive prime numbers with increments equal to 3; that is, can there exist three numbers n, $n + 3$, and $n + 6$ for which each number is prime?

2. A Fermat number is a number of the form $2^{2^n} + 1$. Find two values for n that make the corresponding Fermat numbers equal to primes.

Lecture Nine—Transcript
The Prime Number Theorem and Riemann

In this lecture we'll close our explorations into analytic number theory by studying its crown jewel, the prime number theorem, which answers the delicate question, can we estimate how many primes there are up to a certain size?

This question continues to be one of the great questions in mathematics and of great interest to mathematicians today. To improve the estimates within this important result, we'll return to Euler's general product formula and come face-to-face with a notoriously difficult open problem known as the *Riemann Hypothesis*—one of the most important unsolved problems in all of mathematics.

The truth of the Riemann Hypothesis would immediately imply a long list of deep mathematical results. While the issues surrounding the Riemann Hypothesis first appear to be totally divorced from our "real world," we'll mention some tantalizing new connections with physics through what are known as *random matrices*. Thus, perhaps some modern ideas from physics might hold the key to unlocking the mysteries of this long-standing open question about the atoms of the natural numbers, namely, the primes.

While a complex proof of the prime number theorem was found in 1896, many number theorists wondered if there was a so-called "elementary" argument. Finally, in 1948 an "elementary" proof was found by two great mathematicians: the prolific Hungarian mathematician Paul Erdös and the influential Norwegian mathematician Atle Selberg. As we'll discover, this mathematical milestone led to one of the most famous disputes in all of number theory history.

We'll close this chapter of our course by describing some modern advances in the study of primes, some of which will involve the concept of arithmetic progressions, which we actually saw earlier.

Finally, we'll describe some famous questions involving prime numbers that have confounded number theorists throughout the ages and remain as mysterious today as they were when they were first considered.

We close our study of the primes by returning to Euclid's theorem stating that there are infinitely many primes, and we proved that for ourselves. The question we now wonder is, can we make this result more precise? A common theme in mathematics—once we have a result, can we somehow sharpen it or improve it?

To allow us to discuss the number of primes up to any given point, we'll write the symbol $\pi(n)$ to denote the number of primes less than or equal to n. Here, $\pi(n)$ has nothing to do with the famous number π from geometry and circles; $\pi(n)$ is the name of a value—the value, meaning the number of primes up to a particular value of n. So it's a symbol that number theorists use.

For example $\pi(25)$ would mean the number of primes there are up to 25, and that would equal 9. So $\pi(25)$ is the number 9. Why? Because there are 9 primes that are less than or equal to 25. They are: 2, 3, 5, 7, 11, 13, 17, 19, and 23; and you can see, there are 9 of them.

Euclid's theorem asserts that $\pi(n)$ approaches infinity as n gets larger and larger. Notice how this is the way to phrase Euclid's result using this $\pi(n)$ symbol, because Euclid proved that there are infinitely many primes, which means that as n gets larger and larger, the $\pi(n)$ should eventually get larger and larger as well, and approach infinity.

We're now led to a very natural question in number theory: Is there a formula for $\pi(n)$? Meaning, is there a formula to give how many primes there are up to any given point? This remains an open question whose answer appears to be totally intractable.

By the late 18th century, French mathematician Adrien-Marie Legendre and the great Gauss made an amazing discovery. They observed that the number of primes less than or equal to n seems to be connected with the so-called "natural logarithm" function. The natural logarithm is the logarithm with a very special base number, e. That's a special number that's approximately 2.71828… .

The value of the natural log of n, for any n, can be found using a scientific calculator just using the button that's called "ln," which stands for "natural log."

A logarithm, in essence, is an exponent, and so for thinking purposes we can view the natural log of a natural number n to be, roughly speaking, twice the number of digits n contains. Look at 1,000,000, for example. How many digits are there in the number 1,000,000?

Well, there are 7 digits. If we take a look at the natural log of 1,000,000, what we actually see is something that's roughly 13.8-something. Notice that 13.8 is roughly twice the number of digits. There are 7 digits, double it, we get 14, and the natural log of 1,000,000 is actually 13.8-something. So, you can see, roughly speaking, it's about that.

This rough intuitive notion is certainly enough for us to see that the natural logarithm function is a very slow-growing function. Notice that when we get to 1,000,000, the value of the function is only around 14, so it's very, very slow growing. That's all that matters for us.

The conjecture of Legendre and Gauss was that as n gets larger and larger, the number of primes up to n gets closer and closer to the function n divided by the natural log of n. We can actually look at some data here to see this for ourselves.

For example, if we let n equal 100, then $\pi(n)$—the number of primes there up to n—equals 25. There are 25 primes that are less than or equal to 100. If we now compute 100 divided by the natural log of 100, we see a number that's approximately 21.7-something. So you can see that 25 and 21.7 are relatively close together. It's a pretty good estimate.

If we go out further, we get a better estimate. For example, if we let n = 1,000,000, what's $\pi(1,000,000)$? That asks how many primes are there up to 1,000,000? It turns out that's exactly 78,498. That's how many primes there are. What happens if we compute the other quantity, 1,000,000 divided by the natural log of 1,000,000? That works out to be 72,382.4-something. So you can see these quantities are actually very close together, especially in view of their size.

How could we see if these two quantities are genuinely approaching each other? Let's think about this for ourselves. If two nonzero numbers are equal, we know that their quotient would equal 1, because a number divided by itself is 1. So we can ask, what if we look at the number of primes up to n—so $\pi(n)$—and look at the ratio between that and this quantity n divided by natural log of n. If that ratio is approaching 1, then we could see that these two quantities are, in fact, approaching each other.

We could try this with the 1,000,000 example. If we take a look at n = 1,000,000, and take a look at the number of primes there are up to 1,000,000, and look at the ratio of that number to 1,000,000 over the natural log of 1,000,000, then we'd see a ratio of 1.0844-something—very, very close to 1. Therefore these quantities are in fact getting closer to each other, it appears.

The great Russian mathematician Pafnuty Chebyshev proved in 1850 that if this ratio of $\pi(n)/(n/\ln(n))$ approaches some number, as n gets larger and larger then that number must equal the number 1. It's a very strange result. It says that if these values get closer and closer together then in fact they're heading toward the same value. So, if the ratio is getting closer to something, then it has to be getting closer to 1. Kind of strange.

Nine years later, in 1859, the great German mathematician Bernhard Riemann introduced a number of revolutionary ideas in his memoir entitled *On the Number of Primes Less Than a Given Magnitude*. Among other things, he showed how this issue is connected with complex numbers. Those are numbers involving the imaginary number i, which equals the square root of -1. Also he connected it with the zeta function that we saw earlier, $\zeta(s)$—that actually appears in Euler's general product formula from the previous lecture. Today the zeta function is actually known as the *Riemann zeta function* in honor of Riemann.

Finally, in 1896, French mathematician Jacques Hadamard and Belgian mathematician Charles de la Vallée-Poussin independently produced a proof of the prime number theorem. In other words, they proved that as n gets larger and larger, the number of primes up to n is approaching n divided by the natural log of n. Or more precisely, as we've just seen, as n gets larger and larger, the ratio $\pi(n)/(n/\ln(n))$ approaches 1.

The prime number theorem implies that as n gets larger and larger, the number of primes gets closer and closer to n over the natural log of n. However, for any particular value of n, $\pi(n)$ is not equal to n divided by natural log of n. There's an error given by the difference between these two numbers. This "error" is often referred to as the *error term*.

A natural question remains. How close are these two numbers—the number of primes up to n, and n over the natural log of n—for larger and larger values of n? In other words, how small is the error term?

In Riemann's famous work, he found a profound connection between the number of primes up to n and that zeta function, $\zeta(s)$, which equals $1 + 1/2^s + 1/3^s + 1/4^s + 1/5^s + \cdots$, and so on—that endless, infinite series.

Riemann's insight was to extend this series, the $\zeta(s)$ series, to allow s to be a complex number—that is, he allowed s to equal a number of the form $x + iy$ for decimal numbers of real numbers x and y, and i equaling the imaginary square root of -1.

What does it mean to have an imaginary exponent? Well, while the answer would certainly take us far off topic here, let me just say that it was Euler himself who made this idea precise. In fact, he derived one of the most amazing formulas in all of mathematics that actually includes such an imaginary power. He proved that e—that special number we saw before—raised to the power π times i, is actually equal to -1. Profound and perplexing.

Notice that by raising a positive number like e, which is about 2.7-something, to an imaginary power we can actually get a negative number as an answer. Very bizarre. Well, in fact, imaginary exponents do have meaning in an abstract sense. I'll just leave it at that.

Riemann studied the complex numbers, s, that were solutions to the following very strange-looking equation: $\zeta(s) = 0$. Or equivalently, but even stranger looking, $1 + 1/2^s + 1/3^s + 1/4^s + 1/5^s + \cdots$, and so forth, forever, equals 0.

If we let s be a natural number like 2, we see that the left-hand side will clearly be some positive number and will never be 0. But as we just saw a moment ago, if s were to be imaginary then some of those terms might in fact be negative, and we might have that sum collapsing to 0.

Riemann proved that all the complex solutions, s, to this equation when written as s equals some decimal number x, plus i times some other decimal number y, would always satisfy the condition that that first x that we add in the solution is always a number that's between 0 and 1. It could be 0, all the way up to 1. He showed that if these bounds could be improved, which means if we could tighten those bounds and restrict the x even more then the error term in the prime number theorem could be dramatically reduced, which is of course what we're after.

The key step in Hadamard's and de la Vallée-Poussin's proof of the prime number theorem was showing that all solutions x, written as $x + iy$—to Riemann's equation—actually satisfied a slightly stricter condition. The x value actually had to be strictly larger than 0 and strictly smaller than 1. So instead of being between 0 and 1, we actually see that it has to be strictly inside that region. It can't equal 0, can't equal 1, and that's how they were able to prove the prime number theorem.

Riemann himself conjectured that *all* the solutions to this particular equation satisfied that $x = 1/2$. So in other words, when we take a look at the interval, all the x's lie right in between. There's no need to move around at all. Everything is right in between. Those x's lie on this point right here at 1/2.

This famous conjecture is now known as the *Riemann Hypothesis*. What if it's true? What if the Riemann Hypothesis is true? Well, if it were true, then we would have a much smaller error term in the prime number theorem. In addition, we would learn an enormous amount about the prime numbers because there are literally hundreds of theorems that actually begin by saying, "Assuming the truth of the Riemann Hypothesis, blah-di blah-di blah." So if we knew the Riemann Hypothesis was true, then we would have some interesting results about the primes as well.

In 1900, the great German mathematician David Hilbert included the Riemann Hypothesis on his list of the 23 most important unsolved questions in mathematics at the dawn of the 20th century. A hundred years later, the Clay Mathematics Institute in Cambridge, Massachusetts, listed it as one of its "Millennium Problems"—a correct and complete answer would result in a prize of $1 million.

Let me just suggest that there might be a lot easier ways of making a million dollars. This question is genuinely hard. In fact, it remains one of the most important open questions in all of mathematics today.

Much work has been done, as you could expect, in this direction, and there's a long list of partial results surrounding the Riemann Hypothesis. There's even computer evidence supporting Riemann's conjecture. From 2001 through the year 2005, a number theory program called "ZetaGrid" verified that the first 100 billion solutions to Riemann's equation had the x value equaling 1/2, as Riemann himself conjectured. Of course this data, however overwhelming, is not a general proof.

Very recently, an intriguing and exciting connection has been made between the Riemann Hypothesis and quantum physics. A new direction toward a possible proof of the Riemann Hypothesis might be through the study of objects called *random matrices*. These are mathematical objects that were originally applied to better understand the quantum behavior of larger atoms in physics. Wouldn't it be wonderful if mathematical models that allowed humankind to better understand the atoms of physics also allowed us to gain insights into the atoms of numbers, namely, the primes?

Let's return now to the prime number theorem, which tells us that as n gets larger and larger, the number of primes up to n is approaching n divided by the natural log of n. The original proof of this theorem was extremely clever and subtle, especially in demonstrating that x value is always strictly smaller than 1 for all the solutions to the Riemann equation.

In 1921, the great British analytic number theorist G. H. Hardy wondered if an "elementary" proof of the prime number theorem could be found. In other words, a proof that only involves very simple properties of functions but in a very, very extraordinarily ingenious manner. So, don't use any high-powered math machinery; use simple machinery, but in a very, very clever way.

The proof of the prime number theorem that we've discussed actually used very sophisticated machinery to prove a fact about the prime numbers.

Twenty-seven years later, in 1948, Paul Erdös announced that he and Atle Selberg found a truly "elementary" proof that only involved basic properties of logarithms. This announcement stunned the mathematical world. Let me say a few words about these two great mathematicians with very different personalities and temperaments. In some sense, they were really the "odd couple" of analytic number theory.

Paul Erdös was a Hungarian mathematician who published well over 1500 mathematical articles with over 500 coauthors—collaborators from around the world. He was a mathematical nomad, if you will, having no true academic affiliation for most of his life. He would travel from institution to institution, he would show up at math departments, and he was very famous for showing up at a math department and just walking in and saying to his mathematician friends, "My brain is open," which means, What are you working on? What are you thinking about? Let's talk math. That's why all those papers were written and why he has so many collaborators.

His prolific and foundational work inspired and fed new branches of mathematics, including areas called *graph theory*, *combinatorial number theory*, and *elementary number theory*. Those last of which we saw earlier in the course.

Atle Selberg was a Norwegian mathematician who spent most of his time and his career at the very prestigious Institute for Advanced Study at Princeton. He produced very profound results in advanced areas of number theory, including the so-called *automorphic forms*, and introduced several new areas of study, including the "Selberg sieve," which is similar to the sieve we discussed, but far more advanced, and also what's called the "Selberg trace formula"—a very, very complicated and important formula.

Their proof the prime number theorem was so sensational that in 1950 Selberg was awarded the Fields Medal, which is the mathematical equivalent of the Nobel Prize. In 1952, Erdös received the Cole Prize, one of the most prestigious prizes in mathematics.

This mathematical partnership was both unusual and extremely tense. In fact, a serious controversy arose over who should receive credit for what part of the proof. Here's a brief account of the soap-opera-esque drama:

In March of 1948, Selberg discovered an important formula involving primes but didn't publish it. Several months later, Selberg shared with Hungarian mathematician Paul Turan an inequality he discovered, now known as the *fundamental formula*. Without Selberg's objection, Turan gave a lecture outlining Selberg's recent work. Erdös, who was in the audience, quickly noticed, and he said, You know, I think that you can derive the fact that the ratio of consecutive primes is actually approaching 1 as we go further and further out, and we can derive that just from this inequality.

In fact, within a few hours, Erdös produced an ingenious proof of this extremely important assertion. A day later, when Erdös shared this news with Selberg himself, Selberg totally was dismissive and responded with, Oh, you must have made a mistake—implying that certainly such a deep result that Erdös claims couldn't follow from this little inequality that Selberg had.

A few days later, Selberg, using his formula involving logarithms and primes, together with Erdös's important result, was able to devise an "elementary" proof of the prime number theorem. The key ingredient, however, was Erdös's theorem.

Erdös suggested that the two collaborate and write a joint paper. However, Selberg suggested that each should write their own papers based on their own work. Erdös found this objectionable since this was genuinely an example where the whole was tremendously greater than the sum of the parts—meaning that this result was so enormous that the individual pieces weren't as dramatic as the actual consequence.

To make matters worse, news started to spread about the amazing breakthrough. Now, remember, Erdös traveled all over the place and knew everybody, but the rumor actually attributed the result to Erdös himself. In fact, there's a story that in the fall of 1948, someone greeted Selberg with the exclamation, Have you heard the exciting news of what Erdös has proven? This did not help the situation.

The two did publish their papers separately and mentioned each other's work. However, they no longer spoke to each other after that point. Finally, 45 years later, in 1993, at a conference in Hungary to honor Erdös's 80th birthday, much to everyone's surprise, Selberg not only attended the conference but also delivered one of the lectures. I happened to be attending this conference and actually spoke at this conference and witnessed this myself. Everyone at the conference thought this was the greatest moment in mathematics at that time, that these two great mathematicians were finally reconciling toward the end of their lives.

Little did we know that at that exact same moment, at a different conference in England, even greater news was about to break, and I'll mention this when we talk about Fermat's Last Theorem in a future lecture.

The prime number theorem gives us an estimate for how many primes there are up to any particular point. In some sense, it tells us something about how the primes are distributed. In view of this result, we can consider other advances regarding the distribution of the primes. To do this, let's recall the idea of an arithmetic progression that we saw earlier.

We start with a number and then continue to add a fixed amount. For example, if we start with 1 and repeatedly add 17, then we'd obtain the arithmetic progression that begins 1, 18, 35, 52, and so forth.

Gauss himself wondered if such arithmetic progressions always contained prime numbers. In this example we have 1, 18, 35, 52, 69, 86, 103—and 103 is the very first number on this list that's actually prime. So we had to go all the way to 103 to find a prime.

In 1837, German mathematician Johann Dirichlet proved that given any natural numbers A and B that don't share any prime numbers in common as factors—so they share no prime factors—then the arithmetic progression we get, which starts A, A + B, A + 2B, A + 3B, A + 4B (we keep adding B each time), and so on—this arithmetic progression will not only contain a prime number but will always contain infinitely many prime numbers. Notice how beautifully this extends Euclid's original theorem about the infinitude of primes.

Euclid proved that in the simple arithmetic progression 1, 2, 3, 4, 5, 6, and so forth, that contains infinitely many primes. Now Dirichlet proves that any arithmetic progression, as long as the A and the B have no common factors, any such thing will contain infinite many primes—a dramatic improvement.

In a related but different direction, suppose that we list all the prime numbers in order. So, 2, 3, 5, 7, 11, 13, 17, 19, and so forth. Within this list, do we see any pieces that form an arithmetic progression? The answer is yes. For example, 3, 5, 7 forms a little arithmetic progression because we just have to add 2 each time to get the next one.

What's the *longest* arithmetic progression found in the list of primes? Recently in 2004, some groundbreaking work was done in this direction by Ben Green and Terence Tao. They proved the astounding result that there are arbitrarily long arithmetic progressions within the list of prime numbers. So, any length you can think of, if we look far enough down, we'll find an arithmetic progression of that length.

In 2006, Tao was awarded the Fields Medal for his spectacular work relating to the primes and his work related to number theory in general.

In spite of all that we know about the prime numbers, there are still far more prime number mysteries yet to be cracked. So, we close our discussion on analytic number theory by mentioning some famous— or should I say infamous—open questions involving the primes.

The first question is called the *Twin Prime Conjecture*, which asks, are there infinitely many pairs of primes that differ by 2? For example: 3 and 5 differ by 2, 5 and 7 differ by 2, 11 and 13 differ by 2. These are known as *twin primes*. Notice that the pair 13, 17 is not a pair of twin primes, because their difference is 4.

The Twin Prime Conjecture states that there are infinitely many twin primes, but the proof actually eludes all of us for now. Notice that this would also extend Euclid's result in a different direction.

Next we turn to what is known as *Goldbach's conjecture*. The Goldbach conjecture asserts that every even number greater than 2 can be written as the sum of two prime numbers. We notice that 6, for example, is $3 + 3$—the sum of two primes. Twelve equals $5 + 7$. Even something like 74 equals $31 + 43$—sum of two primes.

This conjecture was first made by Christian Goldbach in a letter to Euler dated back June 7, 1742. Despite the fact that the conjecture has been shown to hold for all even numbers up to around 3×10^{17}—that's huge—a proof that it holds for *all* even numbers has yet to be found.

Finally, just for fun, I want to mention the gigantic number known as *Skewes number* that I mentioned at the beginning of the course. We've seen that the number of primes up to n—that as a function approaches the function n divided by the natural log of n. However, there's another function called the *logarithmic integral*—which I won't define here, which is written as $\mathrm{Li}(n)$—with the property that the $\pi(n)$ function minus this logarithmic integral function actually approaches 0 as n gets larger and larger. It never actually equals 0; it just gets closer and closer to 0. So since we're subtracting these two quantities, the sign will either be positive or negative, depending upon which is larger.

In 1914, the great British mathematician John Littlewood proved that the quantity $\pi(n) - \mathrm{Li}(n)$ changes from positive to negative infinitely many times. So it keeps toggling back and forth which is bigger, but the difference is heading toward 0. But you wiggle back and forth between positive and negative.

However, for all values up to around 10^{22}, the quantity has been shown to be negative. In other words, the logarithmic integral was always larger than the number of primes for all n up to around 10^{22}. Given Littlewood's result, however, we know that the logarithmic integral must, at some point, become smaller than the number of primes from that point. But when?

It was the South African mathematician Samuel Skewes, in 1933, who proved that assuming the truth of the Riemann Hypothesis—so we're assuming the Riemann Hypothesis—the logarithmic integral must be smaller than the number of primes for some n less than—and here's that enormous number—$10^{10^{10^{34}}}$.

This incredibly large bound is now known as *Skewes number* and was described by G. H. Hardy as, "the largest number which has ever served any definite purpose in mathematics."

The primes continue to capture the imagination of individuals from around the world, who continue to uncover the primes' many mysteries—and mysteries, of course, remain. Erdös once said, referring to Einstein's famous quote, "God may not play dice with the universe, but something strange is going on with prime numbers." I hope you enjoyed thinking about the deep ideas of the primes.

Lecture Ten
Division Algorithm and Modular Arithmetic

Scope:

Divisibility is one of the central pillars of number theory. A number evenly divides into another if the first number is a factor of the second. In fact, our work on factorization and the primes from our sojourn into analytic number theory underscores the importance of divisibility. If a number does *not* evenly divide into another, then there is a nonzero remainder resulting from that division. Here we will study these remainders and discover a new and delicate arithmetic known as *modular arithmetic*. We will begin by revisiting the "long division" we saw in elementary school and its sophisticated number theory counterpart known as the *division algorithm*. By repeated applications of the division algorithm we come upon a method for finding the greatest common divisor of any two natural numbers. This important process is known as the *Euclidean algorithm*. We will introduce the main ideas behind modular arithmetic and make the realization that it simply captures the mathematics of cycles—such as hours in the day, in which we return to the same time every 24 hours, or days in the week, in which we return to the same day every seven days. The study of the arithmetic of remainders is important from a mathematical standpoint and also from a practical one. As an illustration, we will consider the sequence of numbers at the bottom of every bank check, called the bank routing number. We will discover a number theoretic coding scheme that involves the modular arithmetic of remainders. Similar error-checking schemes underlie the zebra-striped bar code known as the Universal Product Code (UPC) that is tattooed on nearly all merchandise.

Outline

I. Long division and the division algorithm.

    A. A return to long division from long ago.

        1. To understand the personality of a natural number n, we have been studying the prime numbers that evenly divide into n. We now focus more generally on the idea of division.

2. Using long division, we can divide 47 by 3 and get the answer (the quotient) of 15 with a remainder of 2.

B. A focus on remainders.
   1. Suppose we wish to divide 51 by 4. Before we begin, we wonder, what are the possible values for the remainder?
   2. In this example, because we are dividing by 4, the possible remainders are 0, 1, 2, and 3.
   3. Four goes into 51 twelve times, and we are left with a remainder of 3. If we divide a number a by b, then the remainder will be a number from 0, 1, 2, ... , $b − 1$.
   4. The remainder equals 0 if and only if b divides evenly into a—that is, when b is a factor of a.

C. The division algorithm.
   1. The formalization of long division is called the *division algorithm*: For natural numbers a and b, if we divide a by b, then there exists a unique quotient q and remainder r satisfying $a = bq + r$ and for which $0 \leq r < b$.
   2. In our previous examples we would write: $47 = 3 \times 15 + 2$ and $51 = 4 \times 12 + 3$. Thus without any further work we conclude that 3 is not a factor of 47 and 4 is not a factor of 51.

II. Repeated division and the Euclidean algorithm.

A. The greatest common divisor.
   1. We now move from the study of the arithmetic personality of a natural number (through factorization into primes) to seeing similarities between the arithmetic personalities of two natural numbers.
   2. Given two numbers, we wish to find the largest number that is a common factor of the two given numbers.
   3. For example, if both numbers are even, then we know they both share the common factor of 2, but perhaps they have even more factors in common.
   4. The largest number that is a common factor to both a and b is called the *greatest common factor* of a and b. This number is also known as the *greatest common divisor* of a and b.
   5. If the two given numbers are small, we can easily find their greatest common factor by factoring each number into primes and picking out all the common factors. For

example, if we wish to find the greatest common factor of 30 and 42, we can just factor each: $30 = 2 \times 3 \times 5$, and $42 = 2 \times 3 \times 7$, and then multiply all the common factors: $2 \times 3 = 6$. Thus 6 is the greatest common factor of 30 and 42.

6. This "divide and conquer" method will always work in theory but is impractical for large numbers.

B. The Euclidean algorithm.

1. For large numbers, the question remains, Is there an easy way to find the greatest common divisor of two numbers? The answer is yes: We convert a very difficult task into many easy tasks.

2. In his *Elements of Geometry*, Euclid describes an algorithm for finding the greatest common divisor of two numbers. This algorithm is perhaps the oldest one ever devised and is now known as the *Euclidean algorithm*.

3. The algorithm is based on an important fact: Whenever we know that $a = bq + r$, then the greatest common factor of a and b equals the greatest common factor of b and r. We can repeatedly apply the division algorithm to "divide and conquer" our way to the greatest common factor of two numbers.

C. Some illuminating illustrations.

1. To illustrate the idea, we apply the Euclidean algorithm to find the greatest common factor of 30 and 42.

2. By the division algorithm we see that $42 = 30 \times 1 + 12$. So the greatest common factor of 42 and 30 is the same as for 30 and 12 (these are smaller numbers). We now apply the division algorithm with 30 and 12: $30 = 12 \times 2 + 6$. Again we have that the greatest common factor of 30 and 12 is the same as the greatest common factor of 12 and 6.

3. Now we can see that the greatest common factor of 12 and 6 is 6. We can see this because 6 is a factor of 12, that is, $12 = 6 \times 2 + 0$. When this repeated division algorithm process generates a remainder of 0, then the *previous* remainder (in this example, 6) is the greatest common factor of the original two numbers. This process is the *Euclidean algorithm*.

4. If we apply the Euclidean algorithm to find the greatest common factor of 217 and 245, then we repeat the division algorithm as we just outlined:

$$245 = 217 \times 1 + 28$$
$$217 = 28 \times 7 + 21$$
$$28 = 21 \times 1 + 7$$
$$21 = 7 \times 3 + 0.$$

5. Once we find a remainder of 0, we find the previous (nonzero) remainder, and that equals the greatest common factor. Thus we see that the greatest common factor of 217 and 245 is 7 (and we can check that $217 = 7 \times 31$ and $245 = 5 \times 7 \times 7$).

D. Relatively prime numbers.

    1. Recall that a natural number greater than 1 is prime if it has no factors other than 1 and itself. We now extend this notion to pairs of numbers.

    2. We say two natural numbers a and b are *relatively prime* if they have no prime factors in common—that is, if their greatest common factor is 1.

    3. For example, 6 and 49 are relatively prime (notice that $6 = 2 \times 3$ and $49 = 7 \times 7$). On the other hand, 12 and 20 are *not* relatively prime because they share the common factor of 4.

    4. If we use the Euclidean algorithm with a and b, then those two numbers are relatively prime if and only if the last nonzero remainder equals 1. For example: $49 = 6 \times 8 + 1$.

    5. If we backward-solve for the remainder of 1, we find that $49 - 6 \times 8 = 1$, and then we see we can always find natural numbers x and y that are solutions to $49x - 6y = 1$ (in this case, $x = 1$ and $y = 8$). In general, if two given numbers a and b are relatively prime, then we can find natural numbers x and y that are solutions to $ax - by = 1$. This little fact will be extremely useful in our discussion of cryptography in Lecture Twelve.

III. Modular arithmetic of remainders.

  A. A world of cycles.

    1. Given two natural numbers, if the division algorithm gives a remainder of 0, then we know that one number divides the other; if the remainder is 1, then we know that the two numbers are relatively prime.

2. We now focus on the remainders we see after we divide. As an illustration, let's consider just the remainders when the numbers 0, 1, 2, 3, 4, 5, 6, and so forth are divided by 4.

   Numbers: 0, 1, 2, 3, 4, 5, 6, 7, 8, 9, 10, 11, 12, 13, 14, …
   Remainder ($\div 4$): 0, 1, 2, 3, 0, 1, 2, 3, 0, 1, 2, 3, 0, 1, 2, …

3. We observe that the remainders cycle through the numbers 0, 1, 2, and 3.

B. The cycles of clock arithmetic.

   1. We can view this cycle as similar to that of a clock. In the previous example, once we arrive at 3, the next number we see brings us back to 0.

   2. This is the same arithmetic as we use to tell time (not only in terms of seconds, minutes, and hours, but also in days of the week and months in the year). For example, on a 12-hour clock, fourteen o'clock is really $14 - 12$, which is two o'clock—we cycle around.

   3. The arithmetic of remainders is known as *modular arithmetic*.

   4. Modular arithmetic corresponds to the arithmetic of cycles we use for time. For example, suppose we are considering division by 4. If a number with a remainder of 1 is added to a number with a remainder of 1, then the sum will have a remainder of 2 (for example, $5 + 9 = 14$, and the remainder when 14 is divided by 4 is 2).

   5. If we again consider division by 4, then when a number with a remainder of 2 is added to a number with a remainder of 3, the sum will have a remainder of 1 since we cycle $2 + 3 = (2 + 1 + 1) + 1 = 0 + 1 = 1$ (as remainders when dividing by 4). (For example, $6 + 11 = 17$, and the remainder when 17 is divided by 4 is 1). We only need to focus on the cycles of the remainders and not the quotients. Thus we are always working with numbers relatively small to the divisor.

   6. We refer to this arithmetic of remainders when dividing by 4 as arithmetic "modulo 4" or "mod 4." So we would say $6 + 11 \equiv 1$ (mod 4). The symbol $\equiv$ is read

"congruent" and means "equal remainders" when divided (in this case) by 4.

C. A deeper understanding of numbers.

1. Although people have been studying remainders since at least the 3rd century, when the Chinese were proving interesting theorems involving the cycles of remainders, modular arithmetic was formally introduced by Gauss in 1801.

2. Modular arithmetic is a powerful tool that allows us to establish subtle divisibility results through the use of basic arithmetic. For example, if the difference of two numbers is congruent to 0 (mod m), then that tells us that those two numbers will have the *same* remainders when each is divided by m, without the need to perform the long division.

3. As an illustration, again we consider arithmetic mod 4 (remainders when we divide by 4). Notice that $103 - 95 = 8 \equiv 0$ (mod 4). So 103 will have the same remainder as 95 when divided by 4 *without ever performing the long division*! We can check that $103 = 4 \times 25 + 3$ and $95 = 4 \times 23 + 3$ and see that both have remainders equal to 3.

IV. Applications of modular arithmetic.

A. Breaking into the bank routing numbers.

1. The bank routing number is the nine-digit number located on the lower left corner of any check. It identifies the bank from which the check is issued.

2. Computers scan the routing number (along with the bank account number that appears to the right of the routing number) and electronically make the appropriate withdrawal. To prevent errors in reading these numbers, encoded in the routing numbers is a check that involves modular arithmetic.

3. To check a bank routing number, we take the nine digits, say A, B, C, D, E, F, G, H, I, and produce the following (strange) number: $7A + 3B + 9C + 7D + 3E + 9F + 7G + 3H + 9I$. We now consider this auxiliary number modulo 10 (that is, consider its remainder when divided by 10).

The remainder should be 0, so if the remainder is *not* 0, we know we have an error.

4. Divisibility by 10 is very easy to check: A number is evenly divisible by 10 (so has remainder 0) if and only if the number ends in a 0.

5. For example, one branch of Citizens Bank has a routing number 036 076 150. If we produce the corresponding auxiliary number: $7 \times 0 + 3 \times 3 + 9 \times 6 + 7 \times 0 + 3 \times 7 + 9 \times 6 + 7 \times 1 + 3 \times 5 + 9 \times 0 = 160$, it has a remainder of 0 when divided by 10 ($160 \equiv 0 \pmod{10}$), which is consistent with a valid routing number.

6. We can also use this code to determine a missing digit in a bank routing number. For example, suppose we were able to only partially read a Williamstown Savings Bank routing number. Suppose we only read 211 872 94[] (here the last digit is blocked from sight).

7. We compute the auxiliary number for this bank as best we can: $7 \times 2 + 3 \times 1 + 9 \times 1 + 7 \times 8 + 3 \times 7 + 9 \times 2 + 7 \times 9 + 3 \times 4 + 9 \times []$, which simplifies to $196 + 9 \times []$. We know this number must have a zero remainder when divided by 10. To have a remainder of 0 (mod 10), we must find a multiple of 9 that ends in a 4. So we need to determine what digit [] when multiplied by 9 will end in a 4. The answer is 6. If we let [] equal 6, then our auxiliary number becomes $196 + 54$, which equals 250, giving us a remainder of 0 when divided by 10 and confirming the last digit of the bank code.

8. This error-detection method is guaranteed to detect most common errors: interchanged digits or a single digit read incorrectly.

B. Similar coding schemes and modular arithmetic.

1. Similar coding schemes using modular arithmetic and remainders are used to check Universal Product Codes (UPCs), ISBNs on books, and even driver's licenses in certain states.

2. We employ modular arithmetic—the mathematics of cycles—every day, both consciously when we make appointments or look at a clock and also unconsciously behind the scenes in our technological world.

    3. Number theory allows us to see and understand those invisible instances with great clarity.

Questions to Consider:

1. We are told that the difference between the numbers 123,456,789 and 213 has a factor of 123. Using this fact, determine the remainder when 123,456,789 is divided by 123.

2. Examine one of your personal checks. Find the bank routing number in the lower left corner and check that it satisfies the formula described in the lecture.

Lecture Ten—Transcript
Division Algorithm and Modular Arithmetic

Divisibility is one of the central pillars of number theory. A number evenly divides into another if the first number is a factor of the second. In fact, our work on factorization and the primes, from our sojourn into analytic number theory, implicitly underscored the important idea of divisibility.

If a number does *not* evenly divide into another, then there is a nonzero remainder resulting from that division. Here, we'll study these remainders and discover a new and delicate arithmetic known as *modular arithmetic*.

We'll begin by revisiting the "long division" we all learned in elementary school and its sophisticated number theory counterpart known as the *division algorithm*. By repeated applications of the division algorithm, we come upon a method for finding the greatest common factor any two natural numbers share. This important process is known as the *Euclidean algorithm*.

As we'll introduce the main ideas behind modular arithmetic, we'll make the realization that this type of arithmetic simply captures the mathematics of cycles, such as hours in the day, in which we return to the exact same time every 24 hours. Or days in the week, in which we return to the same day every seven days. Many individuals refer to modular arithmetic as *clock arithmetic*.

The study of the arithmetic of remainders is important from a mathematical standpoint and also from a practical one. As an illustration, we'll consider the sequence of numbers at the bottom of every bank check, called the bank routing number.

We'll discover that hidden within those digits of debt is a number theoretic coding scheme that reduces the chance of an error in reading the data. This error-checking code actually involves the modular arithmetic of remainders. Thus, we'll realize that every time a check is processed, a number theoretic calculation involving modular arithmetic is taking place right within our checking accounts.

Similar error-checking schemes underlie the zebra-striped bar codes known as the Universal Product Codes—or UPCs—that are tattooed on nearly all merchandise today.

Up to this point in our course, to understand the personality of a natural number n, we've studied the prime numbers that evenly divide into n. In order to identify more subtle personality traits of a natural number, we now turn to the more general idea of division.

Long ago we learned a method for dividing one number into another, called "long division." If we divide 47 by 3, we first notice that 3 goes into 4 once, 1×3 is 3, and $4 - 3$ is 1. We have that number 7, so we bring down the 7 and we see that 3 goes into 17 five times, 5×3 is 15, and when we subtract 15 from 17, we see 2. Two is actually smaller than 3, so the process stops. Thus we conclude that 47 divided by 3 equals 15 with a remainder of 2. The number 15 is called the *quotient*.

Suppose now that we're asked to divide 51 by 4. Before we even embark upon this division challenge, we wonder, what are the possible values for the remainder? In this example, because we're dividing by 4, the possible remainders are 0, 1, 2, and 3. Those are the only possible remainders. Well, 4 goes into 51 twelve times, and we are left with a remainder of 3. In fact, we could check that 51 actually equals $4 \times 12 + 3$, or $48 + 3$, which is 51.

If we divide a number a by a number b, then the remainder will be a number from the list 0, 1, 2, 3, ... , out to $b - 1$. The remainder will equal 0 precisely when b divides evenly into a—in other words, when b is a factor of a.

The formalization of long division is actually called the *division algorithm*, and it states that for natural numbers a and b, if we divide a by b then there exists a unique quotient q and remainder r, satisfying that $a = bq + r$, with the additional condition that the r is between 0 and b. And the r can actually equal 0, but it has to be strictly smaller then b.

In our previous examples, we would write: $47 = 3 \times 15 + 2$. Notice that "+ 2," that's the remainder part, and as we saw already, $51 = 4 \times 12 + 3$. Again, that small number 3 represents the remainder. Thus, since we see in both of these cases nonzero remainders, without any further work we conclude that 3 is not a factor of 47 and 4 is not a factor of 51.

We now move from the study of the arithmetic personality of a single natural number (through factorization into primes) to seeing similarities between the arithmetical personalities of two natural numbers.

Given two natural numbers, we wish to find the largest number that is a common factor of the two given numbers. For example, if both numbers are even, then they would both share the common factor of 2, but perhaps they have still more factors in common.

The largest number that is a common factor to both *a* and *b* is called the *greatest common factor* of *a* and *b*. This number is also known as the *greatest common divisor* of *a* and *b*.

If the two given numbers are small, we can easily find their greatest common factor by just factoring each number into primes and picking out all the common factors. For example, if we wish to find the greatest common factor of 30 and 42, we can just factor each: $30 = 2 \times 3 \times 5$, and $42 = 2 \times 3 \times 7$, and then we just multiply all the common factors. Notice they both share the factor 2, they both share the factor 3, so we multiply 2×3 and get 6. Thus, 6 is the greatest common factor of 30 and 42.

This "divide and conquer" method will always work in theory, but in practice, if we have two huge numbers, it's not very practical. Suppose, for example, we wanted to find the greatest common factor of 217 and 245 without a calculator—not so easy to factor those big numbers quickly, at least not for me.

As we've seen, to find the greatest common factor, we can factor each number and multiply the common factors together. However, if the numbers are large, finding the factors might be very difficult. So the question remains, Is there a practical way to find the greatest common factor of two numbers? The answer is yes, and the way we do this is by realizing we have a challenging task. So instead of performing the challenging task, we convert the challenging task to a long list of very easy tasks—a great life lesson that we learned from number theory, that when faced with a hard, difficult, challenging issue, perhaps it can be broken down into many, many, many pieces, each of which is simple.

In his *Elements of Geometry*, Euclid describes an algorithm for finding the greatest common factor of two numbers. This process is perhaps the oldest algorithm ever devised. This procedure is now known as the *Euclidean algorithm*. The algorithm is based on an important fact, and I want to share the important fact with you right now. Whenever we're given the equation that $a = bq + r$, then the greatest common factor of the a and the b is actually identical to the greatest common factor of the b and the r. So the greatest common factor of the a and the b is the same as the greatest common factor of the b and the r. Just using this fact, we can repeatedly apply the division algorithm to a "divide and conquer" method in order to work our way to the greatest common factor of two numbers.

To illustrate the method, we apply the Euclidean algorithm to find the greatest common factor of 30 and 42. We've already done that. I want to introduce the method with an example that we've already seen.

We begin by dividing 30 into 42. It goes in once with remainder 12. In other words, we see $42 = 30 \times 1 + 12$. So from the principle I just outlined, the greatest common factor of 42 and 30 must be the same for the pair 30 and 12. Notice that (30, 12) is a smaller pair than the original (42, 30). So again, we've reduced things a little bit—just a little bit, though. This is the "divide and conquer" approach.

Therefore, we now divide 12 into 30. It goes in 2 times with a remainder of 6. In other words, $30 = 12 \times 2 + 6$. Again, we have that the greatest common factor of 30 and 12 must be the same as the greatest common factor of 12 and 6. Now we can see that the greatest common factor of 12 and 6 equals 6. We can see this because 6 is a factor of 12. In other words, $12 = 6 \times 2$ plus a remainder of 0.

When this repeated division process generates a remainder of 0, then the *previous* remainder—in this case, 6—is the greatest common factor of the original two numbers. This process is the *Euclidean algorithm*.

If we apply the Euclidean algorithm to find the greatest common factor of 217 and 245, then we just repeat the division process we just outlined. So, now we're going to take a look at how this difficult problem can be broken down into a succession of easy little simple problems.

We divide 217 into 245 and obtain $245 = 217 \times 1$ plus a remainder of 28. So we now divide the 28 into 217. Now we obtain that $217 = 28 \times 7$ plus a remainder of 21. Well, we now divide 21 into 28 and obtain $28 = 21 \times 1$ plus a remainder 7. Now we divide 7 into 21, and this is actually going to be easy to see that $21 = 7 \times 3 + 0$.

We get to 0. So once we find a remainder of 0, we return to the previous nonzero remainder, and that equals the greatest common factor. Thus, we see that the greatest common factor of 217 and 245 is 7. We could check and see if our answer is correct by factoring those numbers, which is a little tricky, but I've worked this out: 217 is 7×31, and $245 = 5 \times 7 \times 7$. Notice the only factor they have in common is 7, so therefore we see that 7 is the greatest common factor of those two numbers. But notice that we found the greatest common factor not by factoring, which is hard, but by just a succession of long divisions, which actually is doable and in fact, in most cases, pretty straightforward and easy.

Recall that a natural number greater than 1 is prime if it has no factors other than 1 and itself. We'll extend this notion to pairs of numbers. We say that two natural numbers a and b are *relatively prime* if they have no prime factors in common. In other words, if their greatest common factor is the number 1.

For example, 6 and 49 are relatively prime. Notice that $6 = 2 \times 3$ and 49 is simply 7×7, and they have no factors in common other than just 1. On the other hand, 12 and 20 are *not* relatively prime because they share the common factor of 4.

If we use the Euclidean algorithm with a and b, then those two numbers are relatively prime precisely when the last nonzero remainder that we see equals 1. For example, if we return to the pair 6 and 49, then we see that $49 = 6 \times 8 + 1$, and that "+ 1" tells us that those numbers are relatively prime.

Here's a neat little fact that I want to share with you. If we backward-solve for the remainder of 1, then in this case we'd find that $49 - (6 \times 8) = 1$. In other words, we see that we can find natural numbers—let me call them x and y—that are solutions to a certain equation. The equation in this case would be $49x - 6y = 1$. In this case, we see that the solution would be x equals the natural number 1 and y the natural number 8. If we plug them in, we just saw from using the Euclidian algorithm—the one-step Euclidian algorithm—that we have a solution to this funny-looking equation.

More generally, if two given numbers—let me call them a and b—are relatively prime, then we can always find natural numbers x and y that are solutions to the equation $ax - by = 1$. This little fact will be extremely useful in our discussion of public key cryptography in

Lecture Twelve, so I wanted to mention it and highlight it here, because we'll return to this very important fact about two numbers that are, indeed, relatively prime.

Given two natural numbers, if the division algorithm yields a remainder of 0, then we know that the smaller number divides evenly into the larger. If the remainder is 1, then we just saw that those two numbers are relatively prime. We learned something about how the two numbers interact with each other.

We now focus on the remainders themselves and see after we divide, what we get. As an illustration, let's consider just the remainders when the numbers 0, 1, 2, 3, 4, 5, 6, 7, 8, 9, 10, and so forth are divided by 4. So, let's take a look at the entire list of natural numbers, including 0 now, and divide by 4, and just look at the remainders.

If we take 0 and divide it by 4, we get a remainder of 0. If we take 1 and divide it by 4, we get 0 with a remainder of 1. If we take 2 and divide it by 4, we get 0 with a remainder of 2. If we take 3 and divide it by 4, we see 0 with a remainder of 3. If we take 4 and divide it by 4, we get 1 with a remainder of 0. If we take 5 and divide it by 4, we get 1 with a remainder of 1. Notice what happens; we seem to be cycling. The remainders are 0, 1, 2, 3, 0, 1, 2, 3, and the pattern continues.

We observe that the remainders cycle through the numbers 0, 1, 2, 3, like a clock with four markings on it. For example, the remainders would look like this: We have 0, 1, 2, 3, and what we're seeing is we have a remainder of 1, then a remainder of 2, then a remainder of 3, and then we come back to a remainder of 0 again. Then we have 1, 2, 3, and so forth.

We could actually consider this clock, that has these numbers on it, and we can work from there. We cycle around and around. Notice that once we arrive at 3, then the next number we see is back to 0, just like a clock. This is similar to the arithmetic that is used to tell time, not only in terms of seconds and minutes and hours, but also in days of the week, months in the year, and so forth. For example, on a 12-hour clock, fourteen o'clock is really $14 - 12$, which is two o'clock. We cycle around once, and we pass through that 12 marking.

This type of arithmetic—the arithmetic of remainders—is known as *modular arithmetic.* Modular arithmetic corresponds to the arithmetic of cycles that we use for time, as we just saw. For example, suppose we consider a division by 4 again. If a number having a remainder of 1 is added to a number having a remainder of 1, then the sum will have what remainder? Let's take a look and see.

Here's our clock. Suppose that we have a remainder of 1 and then we add 1. If we add one more remainder, we see that in fact the remainder would be 2. So for example, let's take a look at $5 + 9$, which equals 14, and look at the remainder when 14 is divided by 4, and we see that it's 2. Notice that the remainder when 5 is divided by 4 is 1, and the remainder when 9 is divided by 4 is 1, and so we're seeing $1 + 1 = 2$, and the clock shows that.

If we again consider division by 4, then when a number having a remainder, let's say, of 2 is added to a number having a remainder of 3, let's see what we get. So we start off with a remainder of 2, so here we are, and now we're going to add 3—a remainder of 3. So I add: 1, 2, 3. The remainder should be 1. We can see that on this clock, but we could actually look at an example.

For example, let's just notice that $6 + 11 = 17$. Now when we divide these numbers by 4, what do we see? We see that 6 has a remainder of 2 and 11 has remainder of 3, while 17 has remainder of 1, and so we do see that $2 + 3 = 1$ when we just look at the remainders.

So, we need only focus on the cycle of remainders, and we no longer need to focus on the quotients themselves, if all we're after is the study of remainders. Thus, we're always working with numbers relatively small compared to the divisor, because they're always between 0 and 1 less than the divisor.

We refer to this arithmetic of remainders when dividing by 4 as arithmetic "modulo 4" or, abbreviated, "mod 4." That means that all we're looking at is the remainders when we divide by 4—modulo 4. So the clock has 4 entries from 0 to 3, and then we spin around.

So, given our previous example, we would say—and let me introduce some notation here—what we do is we don't say that $6 + 11 = 1$, but we use this idea of equivalent, and we use, instead of the usual equal sign that has two little horizontal lines, we use three, $\equiv$, we up it, which means that they're not identical, but they have the same remainder. We read the "three equal signs" as either "equivalent" or "congruent."

So we would say that $6 + 11$ is congruent to 1 (mod 4). What it means is that the remainder of $6 + 11$ is actually going to equal 1 when we consider division by 4. Because $5 + 9 = 14$, and 14 has a remainder of 2 when divided by 4, then we could write, using this notation, $5 + 9 \equiv 2$ (mod 4).

Although people have been studying remainders since at least the 3rd century, when the Chinese were proving interesting theorems involving the cycles of remainders, modular arithmetic was formally introduced by Gauss in 1801.

Modular arithmetic is a powerful tool that allows us to establish subtle divisibility results through the use of basic arithmetic, and that's the power. We can use basic tools to say something about subtle issues. For example, if the difference between two numbers is congruent to 0 (mod m), then that tells us that those two numbers will have the *same* remainders when each is divided by m, without the need to actually perform the long division.

As an illustration, we again consider arithmetic mod 4—again, remainders when we divide by 4. Notice that $103 - 95 = 8$, and when you divide by 4, the remainder, when you divide by 4, of 8, is going to be 0. So, $8 \equiv 0$ (mod 4). So, we see that $103 - 95 \equiv 0$ (mod 4). So we conclude that 103 has the same remainder as 95 when divided by 4, and we deduced this fact *without ever performing the long divisions*. You can see the power of this analysis.

In fact, we can check this, in case you have doubts. Using the division algorithm, we see that 103 is actually $4 \times 25 + 3$—that's pretty easy—and 95 is $4 \times 23 + 3$. We've then confirmed that indeed—that these two numbers do have the same remainders. Both remainders are, in fact, equal to 3.

To illustrate the power of modular arithmetic, I want to consider a real-world example. While modular arithmetic plays a central role in mathematics in general, and in number theory in particular, here we offer some illustrations of how modular arithmetic is used in our everyday lives.

The bank routing number is the nine-digit number located on the lower left corner of any check, and in fact you might want to run, get out your checkbooks, and follow along. It identifies the bank from which the check is issued. Each bank will have a different routing number.

Computers scan the routing number, along with the bank account number that appears to the right of the routing number, and electronically make appropriate withdrawals. To prevent errors in reading these numbers, encoded in the routing numbers is a check that involves modular arithmetic.

To check a bank routing number, we take the nine digits that we see, and I'm going to give them names now. Let me call them A, B, C, D, E, F, G, H, I. Each of those represents one of the digits in this nine-digit bank routing number. We produce the following very strange combination of these digits.

We're going to take the first three, A, B, C, and we'll multiply the first number, A, by 7; the next number, B, by 3; and that last number, C, by 9. Then we're going to look at the next three numbers, D, E, F, and do the same thing: Multiply the first by 7, the next by 3, the last by 9, and then with the last three digits, G, H, I, we repeat: 7, 3, and 9. Then we take all those numbers, by the way, and I want to add them all together.

So here is the strange number we concoct: $7A + 3B + 9C + 7D + 3E + 9F + 7G + 3H + 9I$. A little peculiar looking, but you have to admit, it's just a number, and pretty easy to figure out.

We now consider this auxiliary number modulo 10—that is, we consider its remainder when divided by 10. The bank routing number is designed so that this remainder equals 0. So if the remainder is *not* 0, we know we have an error in reading the number.

Divisibility by 10 is very easy to check: A number is evenly divisible by 10—so it has a remainder 0—precisely when the number ends in a 0. For example, one branch of Citizens Bank has routing number 036 076 150. If we produce the corresponding auxiliary number, we'd see $7 \times 0 + 3 \times 3 + 9 \times 6 + 7 \times 0 + 3 \times 7 + 9 \times 6 + 7 \times 1 + 3 \times 5 + 9 \times 0$, and that works out to be 160. I worked that out in advance, actually. I didn't do that in my head.

Well, 160, notice, has a remainder of 0 when divided by 10 because that last digit is a 0. Notice that I'd write this as $160 \equiv 0 \pmod{10}$, which is consistent with a valid routing number.

We can use this code to determine a missing digit in a bank routing number. For example, suppose we were able to only partially read a Williamstown Savings Bank routing number. Specifically, suppose that we knew only the first few numbers—in particular, let's say that we read 211 872 94[], and that last digit was somehow not read, so that last digit's blocked from sight.

We can compute the auxiliary number for this bank number as best as we can: $7 \times 2 + 3 \times 1 + 9 \times 1 + 7 \times 8 + 3 \times 7 + 9 \times 2 + 7 \times 9 + 3 \times 4 + 9 \times$ unknown—so $9 \times$ []—which simplifies down, if you do all the addition that you can, to $196 + 9 \times$ [], the unknown digit.

We know that this number must have a zero remainder when divided by 10. To have a remainder of 0 (mod 10), we must find a multiple of 9—that mystery number when multiplied by 9 has to end in a 4, so that that digit 4, when added to the last digit of the other number, 6, will produce a number that ends in 0.

So, we need to think to ourselves and determine what digit, when multiplied by 9, will end in a 4? The answer is 6, because 9×6 is 54. So if we let our mysterious digit equal 6, then our auxiliary number becomes $196 + 54$, which equals 250, which indeed has a remainder of 0 when divided by 10. Thus we figured out the last digit of this bank code—it ends in a 6—and we did that using modular arithmetic.

In fact, you could try this at parties if you want. Have someone rip out a check and cover up one of the digits. You have to memorize that special way of producing the auxiliary number, but then if you can do the arithmetic in your head or even not, in fact, you can then tell what the missing digit is. Hopefully, they'll let you keep the check.

This error-detection method is guaranteed to detect most common errors in reading bank routing numbers, including interchanged digits and a single digit read incorrectly.

Similar coding schemes using modular arithmetic and remainders are used to check Universal Product Codes—UPCs—and ISBN numbers on books, and even certain driver's licenses in certain states.

We employ modular arithmetic—the mathematics of cycles—every day, both consciously when we make appointments or read a clock, but also unconsciously, behind the scenes, in our technological world. Number theory allows us to see and understand many of these invisible instances with great clarity.

Lecture Eleven
Cryptography and Fermat's Little Theorem

Scope:

In this lecture we will combine ideas from the theory of prime numbers and modular arithmetic to develop an extremely powerful, important, and counterintuitive application: public key cryptography. We will open with a brief historical overview of ciphers and the need for encrypting messages. We then will consider the seeming ridiculous question, Is there an encryption method in which everyone can publicly announce the encryption scheme to code messages yet only the receiver can decode the messages? The surprising answer is that such a public key encryption scheme does indeed exist, and it offers a currently unbreakable coding method. In fact, we use this encryption scheme every day. The main theorem behind this modern method is a 350-year-old result due to Pierre de Fermat involving primes and modular arithmetic. The mathematical secret exploited in this coding method is the reality that factoring large natural numbers into primes is, in practice, extremely difficult, if not "impossible."

Outline

I. A brief history of secret ciphers.

    A. A need for sharing secrets.

        1. Throughout history, humankind has had a desire to keep certain information hidden from certain individuals.

        2. Communications involving business transactions, national security, military plans, and even some romantic trysts needed to be kept secret from various parties.

    B. Early ciphers.

        1. The earliest known example of cryptography was found in Egyptian hieroglyphics around 2500 B.C.E. These may have been for amusement rather than secret communication.

        2. The earliest simple substitution ciphers, known as *monoalphabetic substitution ciphers*, may have been those used by Hebrew scholars around 550 B.C.E.

        3. A *Caesar cipher*, named after Julius Caesar, is a special case of a monoalphabetic substitution cipher in which

each letter is replaced by the letter a fixed number of positions down the alphabet. Such monoalphabetic encryption schemes are very easy to break.

4. Cryptography was known in India by the 1st century C.E., during the time of the famous *Kama Sutra*, in which encryption was suggested as a method for secret communication between lovers.

C. Machines that encode and decode.

1. Ancient Greeks used transposition ciphers, in which elements of the text are rearranged according to a particular scheme. They used a tool called a *scytale* to encrypt and decrypt messages.

2. The Jefferson disk, invented by Thomas Jefferson in 1795, was an encryption and decryption device involving circular disks. A variation of the Jefferson disk was used by the United States Army from 1923 through 1942.

3. The idea of using disks and cogs led to one of the most famous encryption devices, known as the *Wehrmacht Enigma*, which was used by the Nazi military before and during World War II. The Enigma was so complex that for some models, the number of possible rotor configurations exceeded 10^{22}. The great British mathematician and computer scientist Alan Turing was a key figure in cracking the Enigma.

D. Breaking codes.

1. Long before Turing and his team broke the Enigma in the 1940s, people were breaking codes.

2. The first systematic work in cryptanalysis may have arisen from an in-depth religious analysis of the Koran around 800 C.E. Arabs developed the method of frequency analysis to break codes.

3. Very effective for monoalphabetic ciphers, the frequency analysis represents what may be the earliest recorded work in probability and statistics.

4. Frequency analysis can be applied in the recreational solving of cryptoquote puzzles in newspapers.

II. Should and must we trust our allies?

    A. A fundamental flaw.

        1. In all these coding schemes, there is a basic reality: We must trust our friends.

        2. Friends know the encryption and decryption methods for sharing confidential information.

        3. Our friends might be totally trustworthy, but if they accidentally lose the coding instructions, then the encryption system's security would be breached.

    B. Reversing the encryption process.

        1. In the ciphers we described, to decode an encrypted message, one reverses the encryption process.

        2. Thus if people know how to encode a message to us, then they can also decode messages.

III. An intuitive look at a public secret code.

    A. A cipher fantasy.

        1. In the best of all possible worlds, we would not have to trust our friends.

        2. If they lose the codebook, it would not jeopardize the coding scheme.

    B. Making encoding both public and private.

        1. Ideally, knowing how to encode a message would not provide any information as to how to decode the message.

        2. If this fantasy were real, then there would be no need to keep the encoding process a guarded secret.

        3. Instructions on how to encode could be made public, and only the decoding process would need to be kept secret.

        4. In fact, in this fantasy, the encoded messages could be made public as well.

        5. Because the public encoding process could be run backwards to decode the message, there is a need for a secret within the public encryption process.

    C. An intuitive insight through number theory.

        1. Here we apply the concepts we have seen so far to show that such a cryptography fantasy can be made a reality.

        2. The main question remains: How can the encryption scheme be at once public (everyone knows how to code messages) and private (only the rightful receiver can decode the messages)?

3. Such ciphers are known as *public key codes.*

4. Combining the concepts of prime numbers together with modular arithmetic in a clever way allows us to make our fantasy a reality.

D. Shuffling messages.

1. We first offer a metaphor that captures the idea of this modern encryption scheme.

2. Suppose we take a brand new deck of 52 playing cards and perform eight perfect shuffles (also known as *faro shuffles*). Then we would have the cards returning to their original order.

3. If we performed five perfect shuffles, then the order of the cards would look thoroughly mixed, without any semblance of pattern or structure. However, we know a systematic method that would return this jumbled mess back into a familiar, less chaotic pattern. We perform three more perfect shuffles and *voilà*—the cards are transformed from a random mess to their original order!

4. We could employ this shuffling idea to produce an encryption scheme. Our friend could write a message to us, one letter on each card, and encode the message by performing n perfect shuffles (a pre-agreed upon number, n).

5. We would receive the shuffled deck and know exactly what to do: We would perform $8 - n$ perfect shuffles to decode the message.

6. To have this scheme truly fulfill our encryption fantasy, we need to figure out how to mathematically "shuffle" our message and then how to make the shuffling process public.

7. The public feature arises from the fact that factoring extremely large natural numbers is *practically* impossible despite the reality that we know that such a factorization is possible *in theory.*

IV. Shuffling numbers with Fermat's Little Theorem.

    A. Pierre de Fermat: the man and his mathematics.

        1. Pierre de Fermat was a 17th-century French lawyer who explored number theory as a leisure activity. He rarely published his scholarly work.

        2. His body of work comes only from his notes and correspondence with mathematicians. He would provide few if any details into the proofs of his assertions. There were and remain mathematicians (including Gauss) who doubt Fermat had complete proofs of all his mathematical assertions. Many of his claims were not proved until 100 years after he died.

        3. One assertion of his was very stubborn; no one was able to prove or disprove it. Since it was the last assertion that remained unverified, it became known as *Fermat's Last Theorem*. We will study this famous question and the solution that was over 350 years in the making in Lectures Fourteen and Eighteen.

        4. Fermat produced an extremely important and useful theorem that holds the key to our cryptography conundrum. To distinguish this result from his "Last Theorem," this result is known as *Fermat's Little Theorem*.

    B. A pattern within the primes.

        1. To explore Fermat's Little Theorem, we will consider all the possible nonzero remainders when dividing by 5. Those remainders are 1, 2, 3, and 4.

        2. We now consider the remainders when we first multiply these four numbers by 2 and divide by 5, and then we repeat the process with the new list. We would see:

	1, 2, 3, 4
× 2:	2, 4, 6, 8
mod 5:	2, 4, 1, 3
× 2:	4, 8, 2, 6
mod 5:	4, 3, 2, 1
× 2:	8, 6, 4, 2
mod 5:	3, 1, 4, 2
× 2:	6, 2, 8, 4
mod 5:	1, 2, 3, 4

3. If we now just focus on the remainders, we first see 1, 2, 3, 4, and then 2, 4, 1, 3, then 4, 3, 2, 1, then 3, 1, 4, 2, and finally 1, 2, 3, 4. Notice this process just shuffles the numbers 1, 2, 3, 4, and we end back where we started.

4. Thus if we focus just on the numbers in the second location, we see 2 then 4 then 3 then 1, then back to 2. That implies that $2 \times 2 \times 2 \times 2$ must have a remainder of 1 when divided by 5.

C. Fermat's Little Theorem.

1. Fermat generalized this last observation. Fermat's Little Theorem: Given a prime number p and any natural number a that is relatively prime to p, then when we divide a^{p-1} by p, the remainder equals 1. Phrased using the congruence notation, we would say $a^{p-1} \equiv 1 \pmod{p}$.

2. As a further illustration, still with the prime $p = 5$, we consider the number 3 and compute $3^4 = 81$, which indeed has a remainder of 1 when divided by 5.

3. In fact, without any calculation at all, we know the remainder when 5 is divided into 777^4. By Fermat's Little Theorem, the remainder equals 1!

4. Moreover, the remainder when 29 is divided into $1{,}000{,}000^{28}$ equals 1!

D. Applications of an ancient theorem.

1. Fermat first stated this result in a letter dated October 18, 1640. He did not include a proof. Instead he wrote, "I would send you the demonstration, if I did not fear it being too long." The first published complete proof is due to Euler from 1736.

2. In the next lecture we will apply this old theorem about primes to the modern technological world of communication.

Questions to Consider:

1. What are some of the important uses of encryption in our day-to-day lives?

2. Without performing any calculations at all, find the remainder of 29 raised to the 30th power, divided by 31.

Lecture Eleven—Transcript
Cryptography and Fermat's Little Theorem

In this lecture we'll combine ideas from the theory of prime numbers and modular arithmetic to develop an extremely powerful, important, and counterintuitive application: public key cryptography.

We'll open with a brief historical overview of ciphers and the need for encrypting messages. Then we'll consider the seemingly ridiculous question, Is there an encryption method in which everyone can publicly announce the encryption scheme to code messages yet only the receiver can decode messages?

The surprising answer is that such a public key encryption scheme does indeed exist, and it offers a currently unbreakable coding method. In fact, the use of this encryption scheme is found every day in electronic communications of sensitive materials.

Here we'll offer an intuitive introduction into the ideas behind this counterintuitive coding scheme and resolve the paradox in which everyone knows the encryption process and yet only the receiver is able to decode the encrypted messages.

The main theme behind this modern method is a 350-year-old result due to Pierre de Fermat, involving primes and modular arithmetic. The mathematical secret exploited in this coding method is the reality that factoring large natural numbers into primes is, in practice, extremely difficult, if not essentially "impossible."

I want to open our discussion with a brief historical overview of secret ciphers. Throughout history, people have had a desire to keep certain information hidden from other people. Communications involving business transactions, national security, military plans, and even some romantic trysts needed to be kept secret from various parties. The earliest known example of cryptography was found in Egyptian hieroglyphics around 2500 B.C.E. These may have been more for amusement than actually for secret communication. The earliest simple substitution ciphers, known as *monoalphabetic substitution ciphers*, may have been used by Hebrew scholars around 550 B.C.E.

In a monoalphabetic substitution cipher, one letter is substituted for another. For example, every occurrence of the letter A might be replaced by the letter W, and all Bs might be replaced by the letter

M, and Cs might get replaced by the letter R, and so forth, down the alphabet. With this particular scheme that I just made up, the word CAB—C-A-B—would be encoded to read "RWM."

A *Caesar cipher* is a special case of the monoalphabetic substitution cipher in which each letter is replaced by the letter a fixed number of positions down the alphabet. For example, if we replace A by C— notice that C is 2 letters away from A—then B would be replaced by 2 letters away from it, which would be D; and C would be replaced by 2 letters away, which would be E; and so forth, all the way down the line, all the way down to Z. When we get to Z, we come back onto the alphabet; so for Z, we go 2 letters later, which would be A, B—it would be B. So Z would be replaced by B.

The Caesar cipher is named after Julius Caesar who, in the 1st century B.C.E., used such a cipher with a shift of 3 to communicate with his generals. Such monoalphabetic encryption schemes are very easy, in fact, to break.

Cryptography was known in India by the 1st century C.E., during the time of the famous *Kama Sutra*, in which encryption was actually suggested as a method for secret communication between lovers. So, here we see a wonderful example where the everyday passion of human beings actually gets melded with the passion of number theory—quite exciting.

Before leaving this historical overview, I wanted to briefly highlight some of the ingenious machines that have been invented to encode and decode messages.

The ancient Greeks used what are called *transposition ciphers*, in which elements of the text are rearranged according to a particular scheme. They actually used a tool called a *scytale* to encrypt and decrypt messages, and I thought I would demonstrate it.

A scytale is basically just a wooden dowel. Sometimes they have flat faces here, but a wooden dowel. And what you would do is you would take either a long strip of parchment or a ribbon or paper, and you would wrap the paper or the ribbon around the dowel. So, if I were to do this, I would see this, and you could see that this thin sheet of paper has been wrapped around. In this case, actually, it's a ribbon that's been wrapped around.

I could actually write my message on here by writing across. Then when I get to the end of the line, just turn a little bit and write more, and turn a little bit and write more, and write my message.

For example, I've actually done this in advance. This is like one of those cooking shows where they do it in advance, but you take a pen and you'd write the message, but if you come here, here's my message. The message says, "Number theory is true beauty." Excellent message. So now the message is written here.

What I would do next, you see, is I would remove the ribbon, which I'm doing live here, and it's not easy to do. Coding is not easy. No one ever said coding was easy, but I'm doing it right now live. Now you would remove it, and when you untangle it, you see, you get this list of letters. In this case, it begins, "R, Y, E, Y, E, R, U, T, B, O, R, U, M," and then we just would write that out as a row, and that would be the encoded message.

A person receives this, and we'd have to agree in advance, by the way, the thickness or the diameter of the scytale. So, then the message is put on a similar ribbon and then wrapped around the scytale, which is the same diameter so the letters then line up. All of a sudden, that peculiar screed of letters turns out to make sense: "Number theory is true beauty." So we see it.

The interesting thing, by the way here, I just want to comment on, is that, notice that we're going around in cycles, so it's almost like doing arithmetic modulo some number—division mod something.

Another wonderful device that I want to share with you is the Jefferson disk. This was an encryption and decryption device involving circular disks invented by Thomas Jefferson in 1795. Jefferson was quite, as you know, a clever person. Variations of the Jefferson disk were actually used by the United States Army from 1923 through 1942.

I wanted to give you a quick little sense of what this is, because it's so wonderfully clever. What we have here are disks, and on each disk what we see are the letters of the alphabet written out in some random order. Each disk has a different random order to them, so just a random collection of disks, a random collection of letters, and so forth.

What you would do is your friend would have the exact same set of disks with the letters appearing in the exact same orders. If we wanted to send a message, let's say you want to send the message, "Math," so I have M, and then I find A, and then I look for T, there's T, and then I need H. Let's see if we can find H. H. So I have M-A-T-H. I put them together, and that spells out "math," right here on this line.

Now what I do is I hold them all together and I just turn all in unison until I get to any particular line, let's say right here: X, S, I, K. That becomes the encrypted message. Our friend gets that, doesn't know what to do with it, so puts in X, S, I, K, on her disks and just turns them around until she finally comes upon a word. "Ah! Math. That must be from Ed." There you have it. So, a really ingenious way of actually producing encryption and decryption, and it is in fact very, very useful.

The idea of using disks and cogs led to one of the most famous encryption devices, known as the *Wehrmacht Enigma,* used by the Nazi military before and during World War II.

The Enigma was so complex that for some of the models, the number of possible rotor configurations actually exceeded 10^{22}—an enormous number. Incredibly, the Allied forces were able to break this seemingly impossible encryption scheme, and the great British mathematician and computer scientist Alan Turing was a key figure in cracking the Enigma.

Long before Turing and his team broke the Enigma in the 1940s, people had been working hard to break secret codes. The first systematic work in code-breaking may have arisen from an in-depth religious analysis of the Koran around 800 C.E.

Arabs developed the method of frequency analysis to break codes. This is where you look for the letters that appear the most often and then gain some information from that. It's a very effective way for monoalphabetic ciphers—breaking them—and represents what may be the earliest recorded work in probability and statistics.

For example, in the English language, we know that the most popular letter is E, and then followed by T, and then A, and O, and I, and N, and so forth. So, if you see a message that's in this monoalphabetic encryption scheme, the most popular letter probably would be an E, and then you could work from there.

Frequency analysis makes monoalphabetic ciphers easy to break for the educated user. A lot of people know this because this technique is often applied in the recreation of solving cryptoquote puzzles in the newspaper. So we've seen these things before.

In all the coding schemes that we've considered, there's a basic reality: We must trust our friends and allies. They know the biggest secret of all—the encryption and decryption methods for sharing confidential information.

Our friends might be totally trustworthy, but if they accidentally lose the coding instructions—say, for example, if the codebook were to fall out of your trusted friend's pocket, or if someone were to break into their house and steal the secret—then the encryption system's security would be breached.

In the basic ciphers that we've considered, to decode an encrypted message, one reverses the encryption process. Thus, if people know how to encode a message to us, then they also have the power to decode other messages.

Wouldn't it be great to have a coding scheme so that when people use it to send us messages, the encoding process is easy for them to use while at the same time, we're absolutely certain that we're the only ones with the ability to decode the messages?

So, in this fantasy cipher, if you will, we wouldn't have to trust our friends at all. If they lose the codebook and it gets into the wrong hands of undesirables, it would not jeopardize the coding scheme at all. In other words, ideally, in our fantasy cipher, knowing how to encode messages would not provide any information as to how to decode the message.

If this fantasy were real, then there would be no need to keep the encoding process a guarded secret. Instructions describing how to encode messages could be made public, and only the decoding process would need to be kept secret. In fact, in this fantasy, the encoded messages themselves could be made public as well. Our friends could take out ads in *The New York Times* with an encrypted message directed to us. Everyone would see it, but we'd be the only people that would know how to decode it.

The problem is, if a nemesis of ours sees a secret message sent, let's say in *The New York Times*, why couldn't he take the encryption process—which we ourselves made public—and just run that process backwards to decode the message made just for us? This is a problem, and to make this fantasy a reality, we would need to have a secret hidden within the public encryption process. So, even though we make this process public, out there in this public announcement, there's something secret.

We're now ready to apply the number theoretic concepts that we've seen so far to show that just such a crypto-fantasy can be a reality. The main question remains: How can the encryption scheme be at once public—everyone knows how to encode messages—and private—only the rightful receiver can decode the messages? Such amazing ciphers are known as *public key codes*, because the key for encryption is made public.

We'll make our fantasy a reality by combining the concepts of prime numbers together with modular arithmetic in an extremely clever and elegant way.

In the next lecture, we'll offer the specific number theoretic details that make this amazing encryption scheme. But for now, I want to offer an intuitive introduction to the ideas behind the fantasy encryption method so that we can reconcile the seemingly paradoxical scenario in which things are at once public and private.

We'll begin with a metaphor that captures the idea of this modern encryption scheme. Let's suppose that we take a brand-new deck of 52 playing cards—so here, by the way, is a bunch of playing cards, and you could see that it's a brand-new deck because they're beautifully in order. There they are in order, you can see them all, like little soldiers lined up—52 of them. There's the ace and so forth, all the way down the line.

Now, if I were to take this deck and perform eight perfect shuffles, which are also known as *faro shuffles*, that means that I cut the deck exactly in half, 26 and 26, and then perfectly shuffle one, the other; one, the other; and alternate without messing up at all—without making a mistake. If I do that eight times, make eight perfect shuffles, so one, two, three, and so forth, down to eight, then if I look at the cards, magically, they'll actually return to their original order. It's absolutely amazing, in fact, and I urge you to try this for yourself, but if you try this, you have to be able to perform eight perfect shuffles in a row.

Suppose now that we performed just five perfect shuffles, then the order of the cards would look thoroughly mixed up, without any semblance of pattern or structure. However, we know a systematic method that would return this jumbled mess back to a familiar, less chaotic pattern. We'd just perform three more shuffles, bringing the number of shuffles up to eight, and *voilà*—the cards are transformed from a random mess back to their original order.

Notice that we could employ this shuffling idea to produce an encryption scheme. Our friend could write her message to us one letter on each card, so she could say, M, A, T, H, and so forth, and write the message. Then she would just shuffle—perfect shuffle—a certain pre-agreed amount of times. For example, let's say five.

So, she performs five perfect shuffles, and then she delivers the deck of cards to us. If anyone looks at them, it just looks all jumbled, but we know exactly what to do. We would shuffle perfectly $8 - 5$, or 3 times—and then we would be able to read the message that she sent us.

Of course, if we were to use this encryption scheme, then anyone sending us a coded message could decode any other message sent to us as easily as we could. Easy, assuming that we can do perfect shuffles.

To have such a scheme truly fulfill our encryption fantasy, we would need to first figure out how to mathematically shuffle our message like a deck of cards and then figure out how to make that shuffling process public without allowing others to unshuffle our message.

Here's the moment where we introduce our number theory. The public feature arises from the fact that factoring extremely large natural numbers is impossible, for all practical purposes, despite the reality that we know that such a factorization is possible *in theory*. So, now we're going to start to make a distinction between *practice* and *theory*.

To see the basic idea behind this public versus secret dichotomy, suppose that someone announced the number 6 and also revealed a secret. The secret is that this number is the product of exactly two primes.

Can we uncover the secret? Well, of course: $6 = 2 \times 3$. There. In some sense, we just kind of broke the code. What if, instead of 6, the announced number that's the product of two primes was 91? Can we break this code?

With some thought, maybe a little bit of arithmetic, we could figure out that 91 actually is 7×13, and thus we broke this code as well, although it took us a little bit longer.

What if the announced number was 2911? Can we break this code? No. Not so easy. But if we use a calculator or a computer, then we'd be able to discover that 2911 actually equals 41×71, and we broke that code, too.

What if the announced number was a 100-digit number? For all practical purposes, even knowing that this number is, in fact, a product of exactly two primes, we personally would have no way of determining what the two factors are. In fact, even computers have limits to the size of numbers that they can actually factor. Factoring is hard even for computers. So this would be a problem, even for computers.

In this way, notice that we can both announce a piece of information publicly—namely, this enormous number—and yet, from a practical point of view, within that public information is a secret that only we, as the receiver, know.

This reality is how individuals will be able to announce an encryption process without revealing the decryption process. To encrypt messages, people will need only use the huge natural number. However, to decrypt or decode an encoded message, the receiver will need the prime factors of that huge number, which for all practical purposes, is a true secret.

Instead of writing our messages on playing cards, we'll covert our message to numbers in a very simple way. We'll replace the letter A by the number 01, the letter B by the number 02, and so forth, so every letter in the alphabet will be replaced by a two-digit number, even all the way out to Z, which we replace by the number 26.

In order to shuffle our messages, we won't shuffle playing cards; instead we'll shuffle numbers by use of an important number theory theorem due to Fermat.

Pierre de Fermat was a 17th-century French lawyer who explored number theory as a leisure activity. As a result, he rarely published his important scholarly work. His body of work comes to us from his personal notes and correspondence with other mathematicians. As we've noted in a previous lecture, he would provide very little, if any, details into the proofs of his assertions.

There were, and remain, mathematicians—including Gauss himself—who doubted if Fermat had complete proofs of all his mathematical assertions. Many of his claims were not resolved until hundreds of years after his death. Still, Fermat's contributions to number theory were enormous.

One of Fermat's assertions was extremely stubborn in that no one was able to prove or disprove it. Since it was the last assertion of Fermat's that remained unverified, it became known as *Fermat's Last Theorem*. We'll study this famous question and its eventual solution that required over 350 years in the making. We'll discover those ideas in Lectures Fourteen and Eighteen.

From the point of view of ciphers, Fermat produced an extremely important and useful theorem involving the prime numbers that holds the key to our cryptography conundrum. So as to distinguish this important result from his so-called "Last Theorem," this result is known as *Fermat's Little Theorem*.

Informally, Fermat's Little Theorem asserts that if a number having a particular form is divided by a prime, then the remainder will always equal 1.

To inspire the precise statement of Fermat's Little Theorem, we'll consider all the possible nonzero remainders when dividing by the prime number 5. So those nonzero remainders would be 1, 2, 3, and 4.

We'll now consider the remainders when we first multiply these four numbers by 2 and then divide by 5, and then we'll repeat the process with the new list that we get. So here's what we'd see.

We start with 1, 2, 3, 4; we multiply them all by 2, so we see 2, 4, 6, 8. Now we consider their remainders when we divide by 5. In other words, we consider these numbers mod 5, to use the language of the previous lecture. And we'd see that the remainder when 2 is divided by 5 is 2; the remainder when 4 is divided by 5 is 4; the remainder when 6 is divided by 5 is actually 1—we have a remainder of 1—and when 8 is divided by 5, the remainder is 3.

So our list 1, 2, 3, 4 has been transformed into 2, 4, 1, 3, which, notice, is just a scrambling of those numbers. Let's repeat the process. Take this list and multiply by 2, so we get 4, 8, 2, 6. When I consider the remainders when we divide by 5, we would see 4, 3, 2, 1. Now we repeat. We multiply that by 2 and we see 8, 6, 4, 2. When we look at the remainders now—when we divide by 5—we see 3, 1, 4, 2, which notice, again, is just a scrambling up of 1, 2, 3, 4. Then if we do it again, we see 6—doubling—we see 6, 2, 8, 4, which, if we look at the remainders when we divide by 5, we see magically 1, 2, 3, 4—namely, these shuffling of numbers returned us back.

If we now just focus on the list of remainders we just found each time, and, forgetting about the doubling process, we first see 1, 2, 3, 4; and then we see the shuffle 2, 4, 1, 3; then we shuffle again and see 4, 3, 2, 1. Then we shuffle yet again and see 3, 1, 4, 2; and finally, and magically, we return back to 1, 2, 3, 4. Notice how this process does genuinely shuffle the numbers 1, 2, 3, 4 until we return back to the original order, exactly like shuffling the deck of cards eight times.

Since we're multiplying by 2 repeatedly—that's what I decided to look at in this example—let's focus on the values in the second location. Those numbers represent the remainders when different powers of 2 are divided by 5.

Those remainders are 2, then 4, then 3, then 1, and then back to 2. In other words, $2 \times 2 \times 2 \times 2$—or simply 2^4—must have a remainder of 1 when divided by 5. So 2^4 has a remainder of 1 when we divide by 5.

Fermat generalized this last observation, and Fermat's Little Theorem states that given a prime number p and any natural number a that's relatively prime to p—so p doesn't divide into a—then when we divide a raised to the $(p - 1)$ power, and we divide that by p, the remainder equals 1. We could actually phrase this in the language of congruence and use the congruence notation that we introduced, and we'd say that $a^{p-1} \equiv 1$ (mod p), which means that the remainder when we take a^{p-1} and divide it by p equals 1.

In our previous example we found that 2 raised to the power $(5 - 1)$—which is 4—is congruent to 1 (mod 5).

As a further illustration, still with the prime $p = 5$, we consider the number 3 and compute 3^4, which we can actually do. That's $3 \times 3 \times 3 \times 3$, which is 9×9, or 81. If we look at 81, we can actually see that that has a remainder of 1 when divided by 5.

In fact, without any calculation at all, we can instantly produce the remainder when 5 is divided into 777^4. Let's think about this. Let's look at the number 777. Is that relatively prime to 5? Well, if it has a factor of 5 in it, then it has to end in either a 0 or a 5. This ends in a 7, so we know that this number is relatively prime to 5. So we can apply Fermat's theorem, since I'm raising 777 to the $(5 - 1)$, or 4th power. So the remainder, I could automatically announce, equals 1 when divided by 5.

In other words, no division was required; we were able to say the remainder. Moreover, the remainder when the prime 29, say, is divided into $1,000,000^{28}$ must equal 1 for the exact same reason. The only primes that divide into 1,000,000 are 2 and 5, and since the prime 29 doesn't divide into it, it's relatively prime. If I raise 1,000,000 to the $(29 - 1)$, which is 28th, power, I automatically know from Fermat's Little Theorem the remainder is 1.

Fermat first stated this result in a letter dated October 18, 1640. As usual, he did not include a proof. Instead he wrote the following, which I want to share with you. He writes, "I would send you the demonstration, if I did not fear it being too long."

We'll see this is actually a recurring theme. When we come back to Fermat's Last Theorem we'll hear a similar, although slightly more poetic, version of this.

Anyway, the first published complete proof of this theorem is actually due to Euler from 1736. So it took a while to actually nail down the proof.

In the next lecture, we'll apply this historic theorem about primes to the modern technological world of communication. It may seem a bit mysterious right now as to how we'll apply Fermat's Little Theorem to generate an encryption scheme. But I want to not only celebrate that in the next lecture but also notice that we're going to pull almost every topic from number theory that we've explored together, and pull them all into one collective whole, to form this wonderful, interesting application that drives our technological age.

Lecture Twelve
The RSA Encryption Scheme

Scope:

We open this lecture by celebrating the theorem, more than 350 years old, known as *Fermat's Little Theorem*, which connects the primes with modular arithmetic and whose utility permeates throughout all of number theory. It is this important number theoretic result that represents the key to unlocking public key cryptography. Here we will introduce and describe the popular RSA encryption scheme. This clever method of creating ciphers is not only the actual encryption scheme used millions of times a day but also holds within it some deep mathematical ideas. This modern reality of encryption brings to light a number of weighty issues, including the value of information, electronic signatures, and the possibility or impossibility of breaking such a code. As we will see, these questions highlight the interplay between the practical world of our modern technological times and the purely abstract, timeless theorems of number theory. The next time we enter our credit card number on an Internet site or use an ATM, we will realize that we are in fact employing some classical theorems from the theory of numbers in a clever and novel manner.

Outline

I. The return of the primes and Fermat's Little Theorem.

    A. Fermat's Little Theorem.

        1. We recall Fermat's Little Theorem: Given a prime number p and any natural number a that is relatively prime to p, then when we divide a^{p-1} by p, the remainder equals 1. (Phrased using the congruence notation, we say $a^{p-1} \equiv 1 \pmod{p}$).

        2. To see the power of this result, suppose we are given the prime number 7919 (and told that it is prime!). Then for any natural number n that is smaller than 7919, we know that the remainder when n^{7918} is divided by 7919 equals 1. For example, if 5862^{7918} were to be divided by 7919, we know the remainder: 1. No actual multiplication or division is required!

B. The big ideas behind the "little" theorem.

1. Let p be a prime number. Then where did that $p - 1$ exponent come from?

2. Recall from the last lecture we considered all the possible nonzero remainders when a number is divided by p. Those remainders are $1, 2, 3, \ldots, p - 1$.

3. How many of these numbers are relatively prime (share no common factors larger than 1) to p? Again, since p is a prime and the remainders are all less than p, we quickly see that each of them is relatively prime to p. Therefore there are $p - 1$ remainders that are relatively prime to p.

4. The fact that there are $p - 1$ remainders all relatively prime to p is the mathematical key to establishing Fermat's Little Theorem in general. The proof of this theorem, which we will not consider here, incorporates generalized notions from the algebra we have learned in school—an advanced area of mathematics known as *abstract algebra*.

C. Euler's extension of Fermat's result.

1. Even though we will not give the complete proof of Fermat's Little Theorem, we can apply the reasoning we just offered for why the exponent in the theorem is $p - 1$ to discover a generalization of Fermat's result.

2. In particular, is there a corresponding theorem in the case in which we divide by a composite number (rather than the prime p)?

3. The answer is yes, and this extension of Fermat's Little Theorem was discovered by Euler. Let n be any natural number (prime or composite). We now consider all the possible nonzero remainders after division by n: $1, 2, 3, \ldots, n - 1$.

4. Next we count how many of those remainders are relatively prime to n. Recall that in the case in which n is prime, all the remainders are relatively prime to n. Let us write r for the number of remainders that are relatively prime to n.

5. Given the above, Euler proved that for any natural number a that is relatively prime to n, the remainder

when a^r is divided by n equals 1. Symbolically, we say: $a^r \equiv 1 \pmod{n}$.

6. This result is known as *Euler's Theorem*. We notice how this result coincides with Fermat's Little Theorem in the case in which n is a prime number.

7. As an example, if we want to apply Euler's Theorem with a divisor of 21, then we must find out how many of the nonzero remainders from 1, 2, ... , 20 are relatively prime to 21. There are 12 such numbers. Therefore for any number a relatively prime to 21, the remainder when a^{12} is divided by 21 equals 1. For example, $(10^{12})^{12} \equiv 1 \pmod{21}$.

II. An introduction into the RSA encryption scheme.

  A. Revealing the "R," "S," and "A" behind RSA.

    1. In 1977, Ron Rivest, Adi Shamir, and Leonard Adleman, all at MIT, announced an encryption scheme involving primes and modular arithmetic. This encryption scheme is now known as "RSA" in honor of these three mathematicians.

    2. A few years earlier, in 1973, British mathematician Clifford Cook created a similar encryption scheme. However, Cook was working for British intelligence, and thus his work remained unknown until it was declassified in 1997. Cook's work was more of theoretic interest rather than a practical cipher since the calculations required exceeded the capacities of computers in the early 1970s.

  B. Setting up an RSA public key code.

    1. We introduce the steps involved in the RSA encryption scheme with an illustration involving small numbers.

    2. First we need to set up the means by which people can encode messages to us. We select two different prime numbers. Here we chose the tiny primes 3 and 7. We multiply them together and find 21.

    3. Next we consider the product of 1 less than each of these two primes: $(3 - 1) \times (7 - 1) = 2 \times 6 = 12$.

    4. We now select any natural number that is relatively prime to 12; here we pick 29. From our discussion on the Euclidean algorithm in Lecture Ten, we recall that

because 29 and 12 are relatively prime, we can find natural numbers x and y that satisfy the equation $29x - 12y = 1$. In this example, we can let $x = 5$ and $y = 12$. So $(29 \times 5) - (12 \times 12) = 145 - 144 = 1$.

5. We announce the numbers 29 and 21 to the entire world (they are the numbers used for encrypting messages to us); we keep the number 5 a secret—we do not reveal this number to *anyone*, even our friends and allies. We destroy all the other numbers.

C. Encoding and sending secret messages.

1. Suppose now that someone wishes to send us a message letting us know that she will be arriving on the "J" train. She first looks up our encryption numbers and, in this simple example, finds 29 and 21. In actuality these numbers would be enormous.

2. She then translates the message "J" into a natural number using the conversion A = 01, B = 02, ... , Z = 26. So we see that J = 10. She is now ready to encode the message "10" to us.

3. To encrypt the number 10, she computes the remainder when 10^{29} is divided by 21 (notice that 29 and 21 are the two numbers we made public to be used in this manner for encryption).

4. It is easy for computers to find this remainder, which in this example equals 19. The number 19 is the encoded version of "J." She then sends us the secret message 19, which she can post where everyone can see it.

D. Receiving and decoding messages.

1. We receive the encoded message "19," and now have to decode it. To do so, we use the public number 21 and the secret number 5 that *no one* knows.

2. We find the remainder when 19^5 is divided by 21. The remainder equals 10, the original message, which we convert to the letter J. We just decoded the message.

E. Why did this decoding process work?

1. To see why this decoding scheme genuinely works without performing any calculations, we consider the encoded and decoded numbers before we divide by 21.

2. The original message was 10. To encode it, our friend considered 10^{29} (the remainder when divided by 21 is the encrypted message). If we now take this number and raise it to our decoding exponent, 5, we would see: $(10^{29})^5 = 10^{29 \times 5}$.

3. We now recall that these numbers were selected so that: $(29 \times 5) - (12 \times 12) = 1$; that is, $29 \times 5 = 1 + (12 \times 12)$. Applying this equality we see: $(10^{29})^5 = 10^{29 \times 5} = 10^{1+12 \times 12} = 10 \times 10^{12 \times 12} = 10 \times (10^{12})^{12}$. The remainder when 10 is divided by 21 is equal to 10. By Euler's Theorem, as we have already seen, the remainder when $(10^{12})^{12}$ is divided by 21 equals 1. So the remainder when the product $10 \times (10^{12})^{12}$ is divided by 21 equals 10×1, which equals 10: the original message!

4. In actual practice, instead of starting with small primes such as 3 and 7 and taking their product to get 21, the two primes used are enormous, and thus their product is larger still—a number so large that factoring it is, for all practical purposes, impossible.

III. RSA in general.

    A. Setting up an encryption key.

        1. To set up an RSA encryption scheme, we first select two (large) different primes p and q. Let us define $m = p \times q$ and $k = (p - 1) \times (q - 1)$. (In our simple example, $p = 3$ and $q = 7$; $m = 21$; and $k = 12$.)

        2. Next we select any (large) natural number that is relatively prime to k; let us call this large natural number e. (So in our example, $e = 29$.)

        3. We now find natural numbers x and y that satisfy $ex - ky = 1$. (In our example, $x = 5$ and $y = 12$.)

        4. We publicly announce the encryption scheme to send us messages: the numbers e and m. The number m, in practice, is so large that no one can factor it.

        5. We keep x a secret from everyone and then destroy all other numbers. We are now ready to receive encrypted messages.

B. Encrypting messages.

    1. Suppose now that our friend wishes to send us a message. She first converts it to a number, let us call it W, that is relatively prime to m and also less than m.

    2. She then computes the remainder when W^e is divided by m. Let us call this remainder C. The number C is the encrypted version of W. She sends us the encrypted message "C."

C. Decrypting messages. If we receive the encrypted message "C," we know to compute the remainder when C^x is divided by m. That remainder will always equal the original W; that is, we have just decrypted the coded message.

D. Does the decoding scheme always work?

    1. It is a theorem that this scheme of decoding will always return the original message, "W."

    2. The proof of this theorem follows the identical steps we used to see why the decoding scheme worked in the specific example we considered.

    3. This RSA scheme and related schemes are the most popular methods of encryption used today in banking, Internet commerce, and secure communication.

IV. Electronic signatures and the value of information.

    A. Signing our messages.

        1. There are some subtle issues in using this RSA system in practice.

        2. Since everyone knows how to encrypt messages to us, how can we be certain that a message we receive asserting that it is from Zach is really from Zach? Perhaps it is a forged message sent by Marcy.

        3. One way to combat this problem is for the sender to include what is called an *electronic signature*.

        4. An electronic signature can be generated using the sender's secret number that no one else knows. This way we can authenticate the authorship of any message we receive.

B. Breaking the code.

    1. Is this code unbreakable? No. If we factor the number m that is known to be a product of exactly two primes, then we can find the secret decoding number.

    2. Is this factorization approach the only way to break the code? This remains an important open question in cryptography and number theory; namely, is breaking the RSA code equivalent to factoring the number m?

    3. From a practical standpoint, even if there does exist some devilishly sneaky and relatively easy way of breaking the RSA code, as long as no one has found it, the coding scheme remains safe.

C. Selecting the prime numbers for RSA.

    1. Even though we have seen that there is no formula known to generate all the primes, there are methods to generate very large primes. Although factoring is very difficult, even for computers, multiplication of enormous numbers is an easy task.

    2. To create an unbreakable RSA scheme, we need only pick two primes so large that no computer on Earth today can factor their product. In practice we probably will not need to use such gigantic numbers.

    3. How large should our primes be? It depends on the value of the information.

    4. Very important information, such as national security memoranda, would warrant extremely large primes.

    5. The date of a surprise birthday party is not quite as important, and thus fewer people might be willing to invest millions of dollars on computing technology to factor the number m and thus break the code. In this case, small primes would certainly suffice.

    6. The idea of placing a value on information was touted by one of the great foremothers of modern computers and computing, Admiral Grace Hopper.

    7. She saw the importance of determining the cost of replacing lost data long before backing up computers and data was in fashion.

    8. She made enormous contributions to computer science, including being one of the architects of the programming language COBOL.

D. Turning a number from small to enormous.

1. From a number theoretic point of view, the number 400,000,000 is insignificant since almost all natural numbers are larger than it.

2. However there is a way of transforming this small number into an enormous one—even in the eyes of number theorists. Just insert a dollar sign in front: $400 million.

3. This was the price paid by Security Dynamics in 1996 to purchase RSA Data Security—the company formed to promote and sell the RSA systems.

4. Moral: It pays to create number theory theorems.

Questions to Consider:

1. Without performing any calculations at all, find the remainder when 7^4 is divided by 12. Check your answer using a calculator.

2. We saw that one way to break the RSA encryption scheme involves factoring very large numbers, which is possible in theory but impossible in practice with current computer capabilities. What other scenarios can you think of, even beyond number theory, in which something is possible in theory but not in practice?

Lecture Twelve—Transcript
The RSA Encryption Scheme

We open this lecture by celebrating the 350-year-old theorem known as *Fermat's Little Theorem* that connects the prime numbers with modular arithmetic in a subtle manner, and whose utility permeates throughout all of number theory. It is this important number theoretic result that represents the key to unlocking public key cryptography. Here we'll introduce and describe the popular RSA encryption scheme.

This clever method of creating ciphers is not only the actual encryption scheme used millions of times a day, but it holds within it some deep mathematical ideas. This modern reality of encryption brings to light a number of weighty issues, including the value of information, electronic signatures, and the possibility or impossibility of breaking such a code.

As we'll see, these questions highlight the interplay between the practical world of our modern technological age and the purely abstract, timeless theorems of number theory. The next time we enter our credit card number on an Internet site or use an ATM, we'll realize that we are in fact employing some classical theorems from the theory of numbers in a clever and novel manner.

We open this lecture by recalling Fermat's celebrated result known as *Fermat's Little Theorem*, which states, Given any prime number p and any relatively prime number a, then when we divide a^{p-1} by p, the remainder equals 1. We can phrase this using the congruence language and notation, and we'd say that $a^{p-1} \equiv 1 \pmod{p}$—it has a remainder of 1 when divided by p, as long as a is relatively prime to p.

To see the power of this result, suppose we are given the prime number 7919, and suppose that we're told that it's prime, which indeed it is. Then for any natural number n that's smaller than 7919, we know that when n is raised to the power 7918, and when that number is divided by 7919, the remainder equals 1. For example, if we divide 7500 raised to the 7918 power by 7919, we know the remainder is 1. No actual multiplication or division is required. Absolutely amazing.

Let's take a closer look at this incredible result. Suppose that p is a prime number, then where did that $p - 1$ exponent in Fermat's Little

Theorem come from? Recall from the last lecture that we considered all the possible nonzero remainders when a number is divided by p. Those remainders, I remind you, are 1, 2, 3, all the way out to $p - 1$.

We now consider a silly question: How many of these numbers are relatively prime to p? In other words, share no common factors larger than 1 with p. Well, since p is a prime, and all the remainders are less than p, we see that each of them is relatively prime to p, because you can't have the p dividing into a number that's smaller than p. Therefore, we see that there are $p - 1$ remainders that are relatively prime to p. In other words, all the nonzero remainders, when dividing by p, all those values are relatively prime to p.

The fact that there are $p - 1$ remainders all relatively prime to p is the mathematical key to establishing Fermat's Little Theorem in general. That's the idea. The proof of this theorem, which we're not going to consider here, actually incorporates some deep extensions of important ideas from high school algebra. These generalized principles reside within an advanced area of mathematics known as *abstract algebra*.

Even though we're not going to give a complete proof of Fermat's Little Theorem here, we can apply the reasoning we just offered for why the exponent in the theorem is $p - 1$ to discover a generalization of Fermat's result. In particular, we now wonder if there is a corresponding theorem when we divide by a composite number rather than the prime number p.

The answer is yes, and this extension of Fermat's Little Theorem was actually discovered by Euler. We start with any natural number n—it could be prime, or it could be composite. We now consider all possible nonzero remainders after we divide by n. Those remainders are 1, 2, 3, all the way out to $n - 1$.

Next we count how many of these remainders are relatively prime to n. Recall that in the case in which n is prime, then all of these nonzero remainders are relatively prime to n.

Let's write r for the number of remainders that are relatively prime to n. So let's immediately go to an example. If n were to equal 6, then the nonzero remainders would be 1, 2, 3, 4, 5—and of those, only the remainders 1 and 5 are relatively prime to 6, because notice that 2 and 3 and 4 actually share factors with 6. So in this case, since there are only 2 of those nonzero remainders that are relatively prime to 6, the r value would be 2.

Euler proved that for any natural number a that is relatively prime to the number n, the remainder when a to the r power is divided by n equals 1. Symbolically, we could say that $a^r \equiv 1 \pmod{n}$.

So for example, if we let n equal 6, we saw that r would equal 2. If we pick a number that's relatively prime to 6, let's say we pick a equals 5, then we look at 5 to the 2 power, or 5^2. Well, 5^2 is 25, and $25 = (6 \times 4) + 1$. So when divided by 6, its remainder equals 1. In other words, 5^2 is congruent to 1 (mod 6), just as Euler showed.

This result is known as *Euler's Theorem,* and we notice how this result actually coincides with Fermat's Little Theorem in the case in which n is a prime number, because when it's prime, then how many of the nonzero remainders are going to be relatively prime to the n? Well, if n is prime, then all of them are, and so the power will be $n - 1$, which we called $p - 1$ because we were assuming that the n was prime to p.

As an example, if we want to apply Euler's Theorem with a divisor of 21, which notice by the way is 3×7, so it's not prime, then we must find out how many of the nonzero remainders from 1, 2, 3, and so forth, out to 20, are relatively prime to 21—meaning they have no factors of either 3 or 7 within them.

We could actually list them all and count, and if we count, we discover that there are 12 such numbers. In fact, the 12 such numbers are, the ones that are relatively prime to 21 are: 1, 2, 4, 5, 8, 10, 11, 13, 16, 17, 19, and 20. All those numbers share no common factors with 21.

The total number of them is 12. Therefore, for any number a that's relatively prime to 21, when we take a and raise it to the 12th power, and that number is divided by 21, the remainder equals 1. For example, 10 raised to the 12th power will be congruent to 1 (mod 21)—its remainder will be 1 when divided by 21.

In fact, if you think about it, since the only factors of 10 are 2 and 5, and 21 has only factors 3 and 7, what we see here is that if we take 10 and raise it to any power, even a really big one, and take that whole quantity and raise it to the 12th, that will be congruent to 1 (mod 21). I want you to remember this little fact, because we're going to see this in a numerical encryption example in just a few moments.

With Fermat's Little Theorem and Euler's generalization in hand, we are now ready to see the workings of the public key encryption scheme known as "RSA."

In 1977, three mathematicians at MIT introduced an encryption scheme involving primes and modular arithmetic. The three mathematicians were Ron Rivest, Adi Shamir, and Leonard Adleman. The letters of their last names read "R," "S," "A," and hence, this encryption scheme is now known as "RSA."

A few years earlier, in 1973, British mathematician Clifford Cook created a similar encryption scheme. However, Cook was working for the British intelligence, and thus his work remained unknown until it was declassified in 1997. Still, Cook's work was more of a theoretical interest rather than a practical cipher since the calculations required for encoding and decoding exceeded the capacities of computers in the first half of the 1970s. It required the advent of stronger and faster and more sophisticated computers to actually make this encryption scheme a reality.

In order to introduce the steps involved in the RSA encryption scheme, we'll consider an illustration involving very small numbers, just to see the idea. We first need to set up the means by which people can encode messages to us. So, we do this just once. It's kind of the initial setup, and then we're good to go.

The way we do this is we select two different prime numbers. Here, in this example, we'll choose tiny primes. We'll choose 3 and 7. We multiply them together, and we see 21. Next we consider the product of 1 less than each of these two primes. In other words, we take $(3 - 1)$ and we multiply it by $(7 - 1)$, or that is 2×6, which equals 12. So that's the next number we compute.

We now select any natural number that's relatively prime to 12. So we need a number that shares no common factors with 12, and in this example, I'm just going to choose 29. From our discussion on the Euclidean algorithm in Lecture Ten, I want us to recall that because 29 and 12 are relatively prime, then we can find natural numbers x and y that satisfy that funny equation we saw—namely, $29x - 12y$, will equal 1. We can always find natural-number solutions to this because the 29 and the 12 are relatively prime. We're going to use that fact right now.

In this example, we can actually check that one answer is $x = 5$ and $y = 12$. They're solutions because if we plug in x equaling 5 and we plug in 12 for y, we see $(29 \times 5) - (12 \times 12)$, which equals $145 - 144$, which equals 1. Excellent.

With these answers of x and y in our heads, here's what we do: We announce the numbers 29 and 21 to the entire world. They are the numbers used for encrypting secret messages to us. The number we keep secret is the number 5—the number we multiplied the 29 by to make that equation yield 1.

We don't reveal this number 5 to *anyone*, not even our friends or allies. So, suppose that the dearest person in the whole world—you're with that person, your special, special close friend; you're in a romantic embrace, and that person whispers into your ear, "Darling, tell me your secret number." You say, "No," because no one knows this number at all.

We destroy all the other numbers that we used to generate these three, because if they got in the hands of anyone, they could actually, potentially, make the code easy to break. We delete all of those, and we just keep these three numbers: one which we keep secret—the number 5—and the other two we announce to the world.

We're now ready to have people send us encoded messages. Suppose that someone wishes to send us a message letting us know that she'll be arriving on the "J" train. Okay. So she wants to send us the message "J"—the letter J.

She first looks up our encryption numbers, and in this simple example, she finds 29 and 21. Those are the numbers that we announced. In actuality, these numbers would be enormous, having literally hundreds of digits, and the process would actually be done by computers.

She then translates the message "J" into a natural number using the conversion scheme that we mentioned in the previous lecture. We replace every letter by a two-digit number: A is 01, B is 02, all the way out to Z being 26. So we have, in this case, that J equals 10.

She's now ready to encode the message 10 to us. Here's what she does. To encode the number 10, she computes the remainder when

10^{29} is divided by 21. Notice that 29 and 21 are the two numbers that we made public to be used—in this manner—for encryption. So, she takes the message 10, she raises it to the 29th power, and then she divides it by 21, and she sees what the remainder is.

It's actually easy for computers to find this remainder, which in this example, by the way, works out to be 19. The number 19 is the encoded version of "J." She then sends us the secret message "19," which she can do by posting it on the Internet for everyone to see, or even taking out an ad in *The New York Times*. Everyone can see it. No problem at all.

So, we receive the encoded message "19" and now have to decode it. To do so, we use the public number 21 and the secret number 5 that *no one* knows. Don't tell anyone 5.

We compute the remainder now when 19^5 is divided by 21. Now 19 to the 5th is so small that we can actually work it out, and it's 2,476,099. When we divide this by 21, we find a quotient of 117,909 with a remainder of (drum roll please ...) 10. So the remainder equals 10, the original message, which we then convert back to the letter J. We just decoded the message.

Okay. But why did this decoding process work? Was it just a coincidence, or is it, in fact, a fact that it will always work? To see why this decoding scheme genuinely works, we'll consider the encoded and decoded numbers before we divide by 21 and find the remainders.

The original message was 10. Now we're going to do a little bit of number crunching here, so I want you to stay with me. There's going to be a little bit of numbers floating around here. That's okay.

The original message was 10. To encode it, our friend considered 10^{29} and then found the remainder when divided by 21 to deduce the encrypted message. But I want to focus on 10^{29}.

If we now take this number, 10^{29}, and raise it to our decoding exponent, the secret exponent 5, then we'd see 10^{29}, all raised to the 5th power. By the laws of exponents this is equal to $10^{29 \times 5}$. We multiply those two exponents together. So we see 29×5 in the exponent.

This is where the magic happens. We now recall that those numbers, 29 and 5, were selected so that they satisfied that funny-looking equation that we kept talking about. In this case, $(29 \times 5) - (12 \times 12) = 1$. If we now solve for the 29×5, we'd see that $29 \times 5 = 1 + (12 \times 12)$. It's a peculiar identity, but notice that it's true.

So, if we apply this equality, we could write 10^{29} raised to the 5th as $10^{29 \times 5}$, which, by this previous equality, equals 10 raised to the power $1 + 12 \times 12$, all as the exponent, which we could break apart and say is just 10×10 raised to the 12×12, because we add the exponents, there's an invisible 1 on the 10 by itself, and when we add that, we get $1 + 12 \times 12$.

In other words, we see that we have $10 \times (10^{12})^{12}$. Now the remainder when 10 is divided by 21 is equal to 10, because it's actually smaller than 21, so it shares no factors at all, and in fact, by just division, we get 0 with a remainder of 10, so that's easy. And by Euler's Theorem, as we've already seen, if you remember in that second example, the remainder when 10 to any power is raised to the 12th, when divided by 21, always equals 1.

In this case, 10^{12} raised to the 12th, when that's divided by 21, the remainder is 1. So, the remainder when the product $10 \times (10^{12})^{12}$ is divided by 21 equals the remainder of just 10, which is 10, times the remainder of that crazy power, which we know is 1: $10 \times 1 = 10$—the original message.

Of course these small numbers are so tiny that it's easy to factor 21 to be 3×7 and thereby actually break the code. However, in practice, instead of starting with the primes 3 and 7 and taking their product to be 21, the two primes used would be enormous and thus their product would be larger still—in fact, a number so large that factoring it would be, for all practical purposes, impossible, even with the aid of computers.

We now extend our example and describe the RSA system in general. I'm just going to follow the exact same steps that we just outlined, but now I'm going to use letters in place of the numbers so you can see how this works in practice.

First, we've got to set it up. So, to set up an RSA encryption scheme, which we just have to do once, we first select two large, different prime numbers, and we'll call them p and q.

We define the number m to be their product, and we define the number k to be the product of one less than each. So, in our simple example, p was 3, q was 7, m equaled their product—21—and k equaled $(3 - 1) \times (7 - 1)$, or 2×6: 12.

Next we select any large natural number that's relatively prime to k. Let's call this large natural number e. So, in our example, we picked e to be 29. We now find natural numbers x and y that satisfy $ex - ky = 1$—that funny equation that we keep talking about. In our example, x equaled 5 and y was 12.

We now publicly announce the encryption scheme that can be used to send us messages: the two numbers e and m. The number m, in practice, can be chosen so large that no computer today can factor it. So, even though people know it's a product of two primes, no one can actually rip that number apart.

We keep x a secret from everybody and then destroy all the other numbers. We're now ready to receive encrypted messages.

Suppose that our friend wishes to send us a message. She first converts it to a number, let's call it W, that's relatively prime to m and is also less than m. She then computes the remainder when W^e is divided by m. Let us call this remainder C. The number C is the encrypted version of W, and that's what she sends us; that's the encrypted message, "C." So it takes W, raises to the e power, and then sees what the remainder is when we divide by m, and that generates the remainder C, which is the encrypted message.

When we receive the encrypted message "C," we know exactly what to do. We compute the remainder when C raised to the x is divided by m. That remainder will always equal the original message, "W." In other words, we have just decoded the encrypted message.

Does this decoding scheme always work? The answer is yes. It is a theorem that this scheme of decoding will always return the original message, W. In fact, the proof of this theorem follows the identical steps we ourselves used to see why the decoding scheme worked in that specific small example we considered.

This RSA scheme and its variations are the most popular methods of encryption used today in banking, Internet commerce, and secure communication.

We close this lecture by considering some subtle issues involving the use of RSA in practice.

Since everyone knows how to encrypt messages to us, how can we be certain that a message we receive that's asserting that it's actually being sent from Zach is really from Zach? Perhaps it's a forged message sent by Marcy. One way to combat this problem is for the sender to include what's called an *electronic signature*.

An electronic signature can be generated using the sender's secret number that no one else knows. This way, we can authenticate the authorship of any message we receive, and this also is used in practice.

We now face another important question. Is this code unbreakable? The answer is no. If we factor the number m—that is, if we know the two factors that actually make up m, then we can actually find and compute for ourselves the secret decoding number.

Is this factorization approach the only way to break the code? This question remains an important open question in cryptography and number theory—namely, is breaking the RSA code equivalent to factoring the large number m? No one knows for sure. But it's interesting.

In this particular example, from a practical standpoint, even if there does exist some devilishly sneaky and relatively easy way of breaking the RSA code, as long as no one's found it, the coding scheme remains safe. So, even though no one knows if such a sneaky way of breaking RSA is possible, as long as we don't know it the answer is there's no sneaky way of breaking it. Kind of a funny, peculiar look at what theorems really mean.

How do we go about selecting the prime numbers for RSA? We have to pick two large prime numbers. Even though we've seen that there's no formula known to generate the primes, there are methods to generate very large primes. So that's what we use to generate these large prime numbers.

It's interesting—although factoring is extremely difficult, even for computers, multiplication of enormous numbers is a relatively easy task. So, once we have these two enormous primes, a computer can easily multiply them to produce the m. It's hard, though, to try to factor it back and find out what those p's and q's were.

In order to create an unbreakable RSA scheme, we need only pick two primes so large that no computer on Earth today can factor their product, and that way we know it's unbreakable. However, in practice, we probably won't need to use such gigantic numbers.

So, how large should our primes be? It actually depends on the value of the information we're sending. Very important information, such as national security memoranda, would warrant extremely large primes. These are extremely sensitive pieces of information, which we don't want anyone to have.

On the other hand, the date of a surprise birthday party is not quite as important as national security. Fewer people might be willing to invest millions of dollars on computer technology to try to factor the number m and break the code and thus crash the party. In this case, small primes would certainly suffice.

The idea of placing a value on information was touted by one of the great foremothers of modern computers and computing, Admiral Grace Hopper. She saw the importance of determining the cost of replacing lost data long before backing up computers and data were in fashion.

Admiral Hopper made enormous contributions to computer science. She helped design some of the earliest computers and was one of the architects of the programming language called COBOL. She's even credited for finding the first computer bug, and I mean literally, the first bug in a computer. Apparently, a moth somehow found its way into one of Hopper's early computing machines and died. The moth's wing got caught between two arms of a relay switch. The moth was removed, and Admiral Hopper taped the moth right into her computer journal and wrote the entry, "First actual case of bug being found."

I wanted to close this lecture with a final remark about bridging the abstract world of number theory with the real world of everyday living. First of all, within the context of number theory, look at what we've done. We've literally pulled almost every idea we've discovered together in this course together in this one wonderful application.

We use the fact that there are infinitely many primes to know that we can never run out of primes to use to make ever-harder codes to break. We talked about how to factor these numbers, and the importance of factorization; about divisibility; about two numbers being relatively prime gives up that wonderful equation that equals 1. Then of course, modular arithmetic, which led us to actually compute divisibility issues and find remainders without actually doing quotients; and then, of course, finally, Fermat's Little Theorem and Euler's generalization.

Everything came together in this wonderful application—absolutely astounding.

Now, from a number theoretic point of view, the number 400,000,000 is insignificant since almost all natural numbers are larger than it, and we saw, actually, in the very beginning of the course that we're not impressed by these tiny little numbers in number theory. However, there is a way of transforming this relatively tiny number into an enormous one, even within the eyes of the number theorist—just insert a dollar sign in front: $400 million. Now you've even got the number theorists' attention.

Well, $400 million was the price paid by Security Dynamics in 1996 to purchase RSA Data Security, the original company formed to promote and sell the RSA systems.

Our moral: It pays to create interesting theorems in number theory. In fact, Fermat discovered his theorem over 350 years ago, and yet, even though he discovered it so long ago, it would be unimaginable to Fermat to have this beautiful abstract result about primes be the key to the security of all modern electronic communications around the globe.

Seemingly abstract ideas that appear on the fringe today might in fact be the central element of our universe tomorrow.

Timeline

B.C.E.

c. 2000–1650 .... Babylonians apply the Pythagorean Theorem to approximate the square root of 2.

c. 540 .... Pythagoras founds his school and proves the Pythagorean Theorem.

c. 540–500 .... Pythagoreans confounded by their proof that irrational numbers exist.

c. 300 .... Euclid presents his axiomatic method for geometry in *Elements*, in which he proves the infinitude of primes, the irrationality of the square root of 2, and the fundamental theorem of arithmetic. He also presents the Euclidean algorithm for finding the greatest common factor of two natural numbers, perhaps the first algorithm ever created.

c. 200 .... Eratosthenes develops his "sieve" for finding prime numbers up to a given value.

C.E.

c. 200–300 .... Early study of cycles of remainders by Chinese mathematicians that foreshadowed the notion of modular arithmetic.

c. 210–290 .... Diophantus writes the first books on algebra in his 13-volume work *Arithmetica*.

1202 .... Fibonacci explicitly describes the Fibonacci sequence.

1570 .... Bombelli translates Diophantus's *Arithmetica* into Latin, the first step toward making his fundamental work on equations accessible to European scholars.

1575 ....Xylander's Latin translation of *Arithmetica* is the first to be published.

1637 ....Fermat asserts what became known as his "Last Theorem" in the margin of his copy of *Arithmetica*.

1640 ....Fermat asserts his Little Theorem in a letter.

1736 ....Euler gives the first complete published proof of Fermat's Little Theorem.

1737 ....Euler establishes his product formula, marking the beginning of modern analytic number theory.

1742 ....Goldbach conjectures that every natural number greater than 4 can be written as the sum of two primes.

c. 1750..............................Euler uses Fermat's method of descent to show that Fermat's Last Theorem is true for $n = 3$.

1770..............................Lagrange proves that the best rational approximations to an irrational number can be obtained from its continued fraction expansion. He also shows that a real number has a periodic continued fraction expansion if and only if the real number is a real quadratic irrational.

c. 1785..............................Eight-year-old Gauss is said to derive a formula for the sum of the first n natural numbers.

c. 1792	Legendre and Gauss conjecture the statement that later became known as the *prime number theorem.*
1801	Gauss introduces modular arithmetic.
c. 1820	Germain makes important progress on Fermat's Last Theorem.
1825	Dirichlet and Legendre use ideas of Germain to prove Fermat's Last Theorem holds for $n = 5$.
1837	Dirichlet proves result on prime numbers appearing in arithmetic progressions.
1839	Lamé claims to have a proof of Fermat's Last Theorem, but it is flawed.
1842	Dirichlet proves his theorem about rational approximations to real numbers. This work and Liouville's work of 1844 mark the dawn of the area of number theory called *Diophantine approximation.*
1843	Binet derives a formula for the nth Fibonacci number. Kummer uses ideas in Lamé's attempt to prove Fermat's Last Theorem to develop groundbreaking work that marks the birth of algebraic number theory.
1844	Liouville constructs the first examples of transcendental numbers. He also proves his theorem about the size of denominators of rational numbers that closely approximate algebraic numbers.
1847	Kummer extends the ideas of Euler, Germain, Dirichlet, Legendre, and

others to prove that Fermat's Last Theorem is true for all "regular prime" exponents.

1850 ....Chebyshev makes progress on a proof of the prime number theorem.

1859 ....Riemann publishes his seminal paper relating the as-yet-unproven prime number theorem to complex numbers and Euler's product formula. He proposes what became known as the now-famous Riemann Hypothesis.

1873 ....Hermite proves that e is transcendental.

1882 ....Lindemann proves that π is transcendental.

1883 ....Lucas invents the popular Tower of Hanoi puzzle, whose solution involves a recurrence sequence.

1884 ....Kronecker proves that the fractional parts of integer multiples of any irrational number are dense in the interval from 0 to 1.

1896 ....Hadamard and Poussin independently prove the prime number theorem.

1900 ....Hilbert poses 23 questions at the Second International Congress of Mathematics in Paris as a challenge for the 20th century. Many have been solved to date; they are considered milestones.

1921 ....Mordell makes several seminal discoveries on the algebraic structure of rational points on elliptic curves.

1933	Skewes number is the largest ever used in a proof (at that time).
c. 1934	Gelfond-Schneider Theorem asserting the transcendence of certain numbers is proved.
1937	Collatz poses his conjecture on the behavior of a particular sequence of numbers.
1949	Erdös and Selberg each publish proofs of the prime number theorem that use only elementary techniques.
1962	Erdös proves that every real number can be written as the sum of two Liouville numbers.
1970	Matiyasevich proves there cannot exist a general algorithm that will determine in a finite number of steps if an arbitrary Diophantine equation has an integer solution, answering one of Hilbert's 23 questions.
1977	Rivest, Shamir, and Adleman create a public key cryptography system known as RSA encryption. Mazur produces several seminal results involving the algebraic structure of rational points on elliptic curves.
1983	Faltings proves the Mordell conjecture and later wins a Fields Medal.
1993	Wiles proves Fermat's Last Theorem, but an error is uncovered.
1994	Wiles completes a correct proof of Fermat's Last Theorem.

Glossary

absolute value: The distance of a real number from zero on the number line.

additive identity: Zero is the additive identity because $a + 0 = a$ for any number a.

additive inverse: The additive inverse of a number a is $-a$, because $a + -a = 0$, the additive identity. For example, the additive inverse of 5 is -5, and the additive inverse of -17 is $-(-17) = 17$.

algebra: The branch of mathematics that studies equations, their solutions, and their underlying structures.

algebraic geometry: The branch of mathematics that combines ideas from algebra and geometry, with many questions motivated by the need to understand integer solutions to equations.

algebraic integers: A generalization of integers involving algebraic numbers such as $4 + 3i$ or $2 + 6\sqrt{5}$.

algebraic number theory: The branch of number theory that studies numbers that are solutions to certain polynomial equations.

algebraic numbers: The collection of all numbers that are solutions to nontrivial polynomials with integer coefficients.

almost all: A portion of a collection is said to be "almost all" of that collection if when an item is selected at random from the entire collection the chance of choosing something inside that portion is mathematically 100%.

analysis: The branch of mathematics that generalizes the ideas from calculus, especially notions of distance and continuous change.

analytic number theory: The branch of number theory that studies the integers (especially primes) using ideas from calculus and analysis.

arithmetic progression: A list of numbers in which the difference between any number and its successor is always the same value.

ascent, Fermat's method of: A method by which one integer solution to a Diophantine equation gives rise to infinitely many integer solutions.

axiom: A fundamental mathematical statement that is accepted as true without rigorous proof.

best rational approximants: The sequence of rational numbers that approximate a given real number as closely as possible in terms of the size of the denominators.

Binet formula: A formula for the nth Fibonacci number:

$$F_n = \frac{(1+\sqrt{5})^n - (1-\sqrt{5})^n}{\sqrt{5}}.$$

calculus: The branch of mathematics that studies continuous processes and instantaneous rates of change based on precise measures of distance.

chromatic scale: A sequence of 12 pitches in Western music that starts at one pitch and increases frequency of each pitch at regular intervals until reaching the pitch an octave higher.

coefficient: In a polynomial, a coefficient is the number multiplied by a power of the unknown. For example, in the polynomial $27x^8 + 7x^3 - 8x$, 27 is the coefficient of x^8.

Collatz conjecture: A conjecture made in 1937 by Lothar Collatz that a particular algebraic process always terminates at the value 1, regardless of the starting value.

complex numbers: The collection of all numbers of the form $x + yi$, where x and y can equal any real number and i is the square root of -1.

complex plane: A representation of the complex numbers, consisting of a plane with horizontal (real) and vertical (imaginary) axes meeting at a right angle at a point called the "origin."

composite number: A natural number greater than 1 that can be written as the product of two smaller natural numbers.

conjecture: A mathematical statement thought to be true but for which a rigorous proof has not yet been found.

continued fraction: A method of writing real numbers as nested fractions within fractions where all the numerators equal 1.

converge: An infinite series is said to converge if the endless sum has a numerical value. A geometric series converges if the ratio of each term to its successor is less than 1 in absolute value.

counting numbers: The collection of numbers 1, 2, 3, 4, 5, and so on. Also called the "natural numbers."

decimal expansion: The representation of a number in base 10. A decimal point separates the places representing 1s, 10s, 100s, and so on, to the left, and the $1/10^{th}$s, $1/100^{th}$s, and so on, to the right.

degree of an algebraic number: The smallest degree possible of a polynomial in an equation for which the algebraic number is a solution.

degree of polynomial: The largest exponent that appears in the polynomial. For example, the polynomial $7x^3 - 5x + 2$ has degree 3.

dense: The rational numbers are dense within the real numbers because between any two distinct real numbers, there is at least one rational number.

descent, Fermat's method of: A method used to show certain Diophantine equations have no integer solutions.

Diophantine approximation: An area of number theory in which one studies how well irrational numbers can be approximated by rational numbers.

Diophantine equation: An equation that involves only addition, subtraction, and multiplication of natural numbers and unknowns, for which integer solutions are sought.

disjoint: Having no elements in common.

distributive law: The arithmetic law that states that $a \times (b + c) = (a \times b) + (a \times c)$, for numbers a, b, and c.

diverge: An infinite series is said to diverge if the endless sum has no meaningful numerical value. The harmonic series $1 + 1/2 + 1/3 + 1/4 + \cdots$ diverges.

division algorithm: A systematized version of "long division" used to find the quotient and remainder when one integer is divided into another.

e: The fundamental parameter in the measure of growth. The value of *e* is 2.71828… and is equal to the limiting value of the expression $(1 + 1/n)^n$ as *n* grows without bound.

elementary number theory: The area of number theory that focuses on fundamental questions about numbers and whose often subtle answers do not involve advanced mathematics.

elliptic curve: A graph of the cubic equation given by $y^2 = x^3 + ax + b$, where *a* and *b* are given integers.

equation: An expression that sets two quantities equal. For example, $2 + 2 = 4$ and $x^2 - 2 = 0$ are equations.

Euclidean algorithm: An algorithm that produces the greatest common factor of two natural numbers. The method involves repeated applications of the division algorithm.

Euler's product formula:

$$\left(\frac{1}{1-1/2}\right) \times \left(\frac{1}{1-1/3}\right) \times \left(\frac{1}{1-1/5}\right) \times \left(\frac{1}{1-1/7}\right) \times \left(\frac{1}{1-1/11}\right) \times \cdots$$
$$= 1 + \frac{1}{2} + \frac{1}{3} + \frac{1}{4} + \frac{1}{5} + \frac{1}{6} + \cdots$$

exponent: A superscript following a number or variable. For example: $2^3 = 2 \times 2 \times 2$. Also called a "power."

factor: A natural number *m* is a factor of an integer *n* if *m* divides evenly into *n*.

Fermat's Last Theorem: Given any fixed natural number exponent *n* greater than 2, there are no natural-number solutions for *x*, *y*, *z* to the equation $x^n + y^n = z^n$.

Fermat's Little Theorem: Given a prime number *p* and any natural number *a* that is relatively prime to *p*, when we divide $a^{(p-1)}$ by *p*, the remainder equals 1.

Fibonacci numbers: The sequence of numbers 1, 1, 2, 3, 5, 8, and so on, in which each number after the first two is equal to the sum of its two predecessors.

Fields Medal: An award given every four years to two to four mathematicians under the age of 40 by the International Mathematics

Union. Considered by many to be the equivalent of a Nobel Prize, which does not exist for math.

fractional part: The decimal part of a real number. For example, the fractional part of 3.14159 is 0.14159.

fundamental theorem of arithmetic: Every natural number greater than 1 can be written uniquely—up to reordering of the factors—as a product of prime numbers.

Gelfond-Schneider Theorem: If an algebraic number not equal to 0 or 1 is raised to an algebraic irrational power, then the result is a transcendental number.

geometric progression: A list of numbers in which any number divided by its successor always gives the same value.

geometric series: The endless sum of all the numbers in a geometric progression.

Germain prime: A prime number p for which the number $2p + 1$ is also prime.

Goldbach conjecture: Goldbach's conjecture states that every even number greater than 4 equals the sum of two primes.

golden ratio: The number $(1+ \sqrt{5})/2$.

greatest common factor: The greatest common factor of two natural numbers is the largest natural number that divides evenly into each of them.

group: A collection of elements and an operation for combining them that satisfies certain special algebraic properties.

harmonic series: The infinite series $1 + 1/2 + 1/3 + 1/4 + 1/5 + \cdots$.

Hilbert's problems: The list of open questions David Hilbert posed at the Congress of Mathematics in 1900. He considered them to be the most important open questions in mathematics for the 20th century.

i: The square root of -1.

ideal: A packet of algebraic integers that exhibits unique factorization into prime ideals.

imaginary numbers: The collection of numbers that are multiples of i.

infinite series: An unending sum of quantities such as $1 + 1/2 + 1/4 + 1/8 + 1/16 + \cdots$. Such a sum may or may not have a numerical value.

integers: The collection of numbers consisting of the natural numbers 1, 2, 3, ... , together with all their negatives and zero.

irrational numbers: The collection of all numbers that are not rational.

Lagrange's Theorem: A real number has a periodic continued fraction expansion if and only if the real number is a real quadratic irrational.

logarithm: The exponent to which a base must be raised to produce a given number. For example, the base-10 logarithm of 1000 is 3, because $10^3 = 1000$.

Lucas sequence: The sequence of numbers 2, 1, 3, 4, 7, 11, and so on, in which each number is the sum of its two predecessors.

multiplicative identity: The number 1 is the multiplicative identity because $1 \times a = a$ for any number a.

multiplicative inverse: The multiplicative inverse of a nonzero number a is its reciprocal, $1/a$, because $a \times 1/a = 1$, the multiplicative identity. For example, the multiplicative inverse of 5 is 1/5, and the multiplicative inverse of 1/2 is 2.

natural numbers: The collection of numbers 1, 2, 3, 4, 5 ...; also called the "counting numbers."

nonrepeating expansion: A number expansion in any base is nonrepeating if it is not periodic.

number line: A representation of the real numbers; a line extending endlessly in both directions, with a point marked as 0 and at least one more point, usually 1, marking the unit of length. Each point on the line corresponds to a real number according to its distance from 0, with points to the right of zero denoting positive numbers and points to the left of zero denoting negative numbers.

number theory: The area of mathematics that focuses on the properties and structure of numbers.

octave: An interval in music where the two pitches have a frequency ratio of 2 to 1.

partial quotients: The natural numbers appearing in a continued fraction expansion.

Pell equation: A Diophantine equation of the form $x^2 - dy^2 = 1$, where d is a given square-free natural number.

perfect fifth: An interval in music in which the two pitches have a frequency ratio of 3 to 2.

periodic expansion: A number expansion in any base is periodic if, eventually, the digits to the right of the decimal point fall into a pattern that repeats forever. Also known as "repeating expansion."

pi: The ratio of the circumference of a circle to its diameter. Pi is denoted by the Greek letter π and equals 3.14159... .

$\pi(n)$: The number of primes less than or equal to the natural number n.

Pigeonhole Principle: The basic but extremely important observation that placing $n + 1$ objects into n pigeonholes results in at least one pigeonhole having two or more objects.

polynomial: An expression involving a single unknown (usually denoted by x), in which various powers of the unknown are multiplied by numbers and then added. For example, $3x^2 - 17x + 5$ and $27x^8 + 7x^3 - 8x$ are polynomials.

prime factorization: Calculation of all the prime factors in a number.

prime number: A natural number greater than 1 that cannot be written as the product of two smaller natural numbers.

prime number theorem: The number of primes less than or equal to a particular natural number n is approximately $ln(n)/n$, where $ln(n)$ denotes the natural logarithm of n. That is, as n increases without bound, the number of primes less than n gets arbitrarily close to $ln(n)/n$.

proof: A sequence of logical assertions, each following from the previous ones, that establishes the truth of a mathematical statement.

public key cryptography: A method of encoding and decoding messages in which the encoding process can be announced publicly.

Pythagorean Theorem: $a^2 + b^2 = c^2$, given a right triangle with side lengths a, b, and c (with c the longest length—the hypotenuse).

Pythagorean triple: A trio of numbers x, y, and z that satisfies the Pythagorean Theorem. The trio 3, 4, 5 is a Pythagorean triple.

ratio: A quantity that compares two measurements by dividing one into the other.

rational numbers: The collection of numbers consisting of all fractions (ratios) of integers with nonzero denominators.

rational point: A point (X, Y) in the coordinate plane for which the X and Y values are both rational numbers.

real numbers: The collection of all decimal numbers, which together make up the real number line.

recurrence sequence: A list of numbers in which, given one or more starting values, subsequent values are produced from preceding values using a given rule.

relatively prime: Two natural numbers are relatively prime if 1 is the largest natural number that divides evenly into both of them; that is, their greatest common factor is 1.

repeating expansion: See "periodic expansion."

Riemann Hypothesis: A conjecture involving the complex number solutions to a particular equation. If true, the Riemann Hypothesis has important implications about the distribution of prime numbers. A prize of $1 million has been offered for a complete proof.

RSA encryption: A popular method of public key cryptography developed by Rivest, Shamir, and Adleman that uses Fermat's Little Theorem.

scytale: A tool used by ancient Greeks to encrypt and decrypt messages.

Sieve of Eratosthenes: A process by which the prime numbers up to a given value can be found.

Skewes number: A number approximately equal to $10^{10^{10^{34}}}$.

slope: The pitch of a line, defined precisely as the ratio of the change in the vertical direction of the line to the change in the horizontal direction.

solution: Given an equation involving an unknown, a number is a solution to the equation if substituting that value for the unknown yields a valid equation.

square root: The square root of a number is a number that when multiplied by itself yields the first number.

square root of 2: $\sqrt{2}$, which equals 1.414... .

square-free number: A natural number whose prime factorization does not contain any particular prime more than once.

theorem: A mathematical statement that has been proven true using rigorous logical reasoning.

Towers of Hanoi: A puzzle consisting of three pegs and a stack of disks of graduated sizes on one peg which are to be transferred to another peg following certain rules.

transcendental numbers: The collection of all numbers that are not algebraic.

triangular numbers: The series of numbers formed by successive sums of the terms of an arithmetic progression, with the first term being 1 and the common difference being 1. The sum $1 + 2 + 3 + \cdots + n$ is the nth triangular number. The first few triangular numbers are 1, 3, 6, 10, 15, and 21.

twin prime conjecture: There are infinitely many twin primes. Two prime numbers are twin primes if their difference is 2.

unique factorization: Every natural number greater than 1 can be written as a product of prime numbers in only one way, up to a reordering of the factors. This product of primes is the unique factorization of the number.

unit circle: The circle of radius 1 centered at the origin; algebraically, it is the collection of all points (X, Y) satisfying the equation $X^2 + Y^2 = 1$.

Biographical Notes

Adleman, Leonard (1945–): This theoretical computer scientist from the University of Southern California studies cryptography and molecular biology. Together with Ronald Rivest and Adi Shamir in 1977, he created a method of public key encryption known as the RSA algorithm.

Bertrand, Joseph (1822–1900): A French mathematician at the École Polytechnique and other schools, Bertrand made contributions in probability and mechanics, as well as his famous conjecture made in 1845 that there is always a prime between n and $2n$.

Binet, Jacques (1786–1856): This French mathematician worked in number theory and matrix theory. The closed form expression for the Fibonacci numbers is named after him.

Cantor, Georg (1845–1918): A German mathematician of Russian heritage and a student of Karl Weierstrass, Cantor established many of the early fundamentals of set theory. Between 1874 and 1884, he created precise ways to compare infinite sets, establishing the existence of infinitely many sizes of infinity, as well as infinitely many irrational and transcendental numbers. The controversy stirred by his work, along with bouts of depression and mental illness, caused him great difficulties later in his life, and he died in a sanatorium.

Chebyshev, Pafnuty (1821–1894): This Russian mathematician proved Bertrand's conjecture that there is always a prime between n and $2n$. His work also contributed to later proofs of the prime number theorem.

Collatz, Lothar (1910–1990): A German mathematician at the University of Hamburg, Collatz may be best known for the Collatz conjecture, which he posed in 1937 and which remains unsolved. He died while attending a math conference.

Diophantus (c. 210–290 C.E.): This Greek mathematician lived in Alexandria, Egypt, where he wrote one of the earliest treatises on solving equations, *Arithmetica*. Although he considered negative numbers to be absurd and did not have a notation for zero, he was one of the first to consider fractions as numbers. In modern number theory, Diophantine analysis is the study of equations with integer coefficients for which integer solutions are sought.

Dirichlet, Johann (1805–1859): A German mathematician at the University of Berlin, Dirichlet made many contributions to number theory, including important work on the distribution of primes and rational approximations. His wife was a sister of composer Felix Mendelssohn.

Eratosthenes (c. 276–194 B.C.E.): A Greek scholar who lived in Egypt, Eratosthenes was a mathematician, geographer, astronomer, and poet. The "sieve" that bears his name is a method for extracting primes from a list of consecutive numbers.

Erdös, Paul (1913–1996): An immensely prolific Hungarian mathematician, Erdös authored or coauthored approximately 1500 papers. Choosing to have no formal professional position, he traveled from institution to institution to work with colleagues in number theory, probability, set theory, combinatorics, and graph theory. He discovered remarkably "elementary" proofs of the Bertrand conjecture and, using a result of Atle Selberg, the prime number theorem.

Euclid (c. 325–265 B.C.E.): The mathematician Euclid lived in Alexandria, Egypt, and his major achievement was *Elements*, a set of 13 books on basic geometry and number theory. His work and style is still fundamental today, and his proofs of the infinitude of primes and the irrationality of $\sqrt{2}$ are considered two of the most elegant arguments in all of mathematics.

Euler, Leonhard (1707–1783): A Swiss mathematician and scientist, Euler was one of the most prolific mathematicians of all time. He introduced standardized notation and contributed unique ideas to all areas of analysis, especially infinite sum formulas for sine, cosine, and e^x. The equation known as Euler's formula, $e^{\pi i} + 1 = 0$, is considered by many to be the most beautiful in all mathematics.

Fermat, Pierre de (1601–1665): This French lawyer was one of the best mathematicians of his time. Sometimes called the creator of modern number theory, Fermat's contributions are numerous, including his so-called "Little Theorem," famous "Last Theorem," and his methods of ascent and descent for analyzing Diophantine equations.

Fibonacci, Leonardo de Pisa (c. 1175–1250): An Italian mathematician, Fibonacci traveled extensively as a merchant in his early life. Perhaps the best mathematician of the 13th century, he introduced the Hindu-Arabic numeral system to Europe and discovered the special sequence of numbers that bears his name.

Gauss, Carl Friedrich (1777–1855): A German mathematician commonly considered the world's best mathematician, Gauss is known as the "Prince of Mathematics." He established mathematical rigor as the standard of proof and provided the first complete proof that complex numbers are algebraically closed, meaning that every polynomial equation with complex coefficients has its solutions among complex numbers.

Germain, Sophie (1776–1831): This French mathematician battled social pressure and discrimination to become one of the most highly regarded women mathematicians of her day. She often wrote using a male pseudonym, but she was admired and mentored by Lagrange and Gauss even more after they discovered she was a woman. She made significant progress on Fermat's Last Theorem and the study of primes, including a particular variety that now bears her name.

Goldbach, Christian (1690–1764): This Prussian-born mathematician is perhaps best known for a conjecture he made in a letter to Euler. The Goldbach conjecture, which claims that every even number greater than 2 can be written as the sum of two primes, is one of the oldest unsolved problems in number theory.

Hadamard, Jacques-Salomon (1865–1963): This French mathematician independently produced a proof of the prime number theorem in 1896, the same year that Charles de la Vallée-Poussin also produced a proof.

Hilbert, David (1862–1943): Born in Prussia, this German mathematician was one of the most broadly accomplished and widely influential in the late 19th century and 20th century. He spent most of his professional life at the University of Göttingen, a top center for mathematical research. His presentation in 1900 of unsolved problems to the International Congress of Mathematics is considered to be one of the most important speeches ever given in mathematics. He was a vocal supporter of Georg Cantor's work and presented the Continuum Hypothesis as the first problem on his list in 1900.

Hopper, Grace Murray (1906–1992): An American mathematician, Hopper was a pioneer in the early days of computer science, writing the first compiler for a computer programming language. She spent most of her career in the Navy and retired at the rank of rear admiral.

Kronecker, Leopold (1823–1891): This German mathematician made contributions in number theory, algebra, and analytic ideas of continuity. As an analyst and logician, he believed that all arithmetic and analysis should be based on the integers and, thus, did not believe in the irrational numbers. This put him at odds with a number of colleagues and, especially, the new ideas of Cantor in the 1870s.

Kummer, Ernst (1810–1893): This German mathematician developed the notion of ideal numbers by exploring the property of unique factorization mistakenly assumed in Lamé's flawed proof of Fermat's Last Theorem.

Lagrange, Joseph-Louis (1736–1813): This French-Italian mathematician was a student of Euler and was considered one of the best mathematicians of his day. He made numerous contributions to number theory, including the theory of continued fractions.

Lamé, Gabriel (1795–1870): This French mathematician made important contributions in number theory. He proved Fermat's Last Theorem for $n = 7$. His clever but flawed attempt at a general proof led to important developments in number theory.

Legendre, Adrien-Marie (1752–1833): This French mathematician made important contributions in number theory, statistics, algebra, and analysis. He proved Fermat's Last Theorem for $n = 5$ independently of and shortly after Dirichlet, and he conjectured the prime number theorem in 1796.

Liouville, Joseph (1809–1882): This French mathematician worked in many fields but is perhaps best known for his proof, given in 1844, of the existence of transcendental numbers. He constructed actual examples and described a special class of transcendental numbers that are now called *Liouville numbers*.

Littlewood, John (1885–1977): This British mathematician made contributions in number theory related to the prime number theorem and the Riemann Hypothesis.

Lucas, Edouard (1842–1891): This French mathematician worked in various areas of number theory, including Diophantine equations.

He studied the Fibonacci sequence extensively, leading to a generalization called *Lucas sequences*. He also invented the Tower of Hanoi puzzle.

Mazur, Barry (1937–): This New York–born mathematician earned his Ph.D. from Princeton at the age of 21 and has been at Harvard for nearly 50 years. He is active in research and teaching and has made important contributions on elliptic curves and other areas of number theory and mathematics.

Mordell, Louis (1888–1972): Born in Philadelphia, this British mathematician worked in number theory, advancing knowledge of rational points on elliptic curves.

Pell, John (1611–1685): This English mathematician studied Diophantine equations, though not the particular type that bears his name, which was mistakenly attributed to Pell by Gauss.

Pythagoras (c. 569–507 B.C.E.): Although best known for the theorem about right triangles that bears his name, Pythagoras had a much broader influence on mathematics and scholarship in general. Born on the Greek island of Samos, he moved to what is now southern Italy and founded a religious and scholarly community called the Brotherhood. Because they left no written records, knowledge about the "Pythagoreans" of the Brotherhood comes from later sources, including Plato and Aristotle. The Brotherhood considered numbers the basis of all reality; Pythagoras is called the "Father of Number Theory."

Ramanujan, Srinivasa (1887–1920): Born and raised in India, Ramanujan received almost no formal training in mathematics and yet is considered one of the mathematical geniuses of the 20th century. He made incredible contributions to number theory and analysis.

Riemann, Bernhard (1826–1866): A major figure in mathematics during the mid-19th century, Riemann made important contributions to analysis, geometry, and topology. Calculus students everywhere know of the Riemann integral. His Ph.D. advisor was Gauss, and he spent his brief career at the University of Göttingen. His conjecture about the distribution of primes, called the *Riemann Hypothesis*, is one of the most important unsolved questions in mathematics today.

Rivest, Ronald (1947–): This computer scientist from MIT works in cryptography. Together with Leonard Adleman and Adi Shamir, he created a method of public key encryption known as the RSA algorithm.

Selberg, Atle (1917–2007): This Norwegian mathematician won the Fields Medal in 1950 for his work relating analysis to the Riemann zeta function. He also discovered an "elementary" proof of the prime number theorem.

Shamir, Adi (1952–): This cryptographer from the Weizmann Institute in Israel created in 1977, together with Leonard Adleman and Ronald Rivest, a method of public key encryption known as the RSA algorithm.

Skewes, Stanley (1899–1988): This South African mathematician is best known for discovering Skewes number in 1933. This number gave a bound on an interval during which values related to prime numbers exhibit certain properties. At the time, this number was the largest number to have significance in a mathematical proof.

Turing, Alan (1912–1954): This British mathematician and cryptographer is considered by many to be the father of modern computer science. During World War II, his work with British intelligence was critical to cracking the Enigma encryption scheme used by the Nazis. In 1966, the Association for Computing Machinery established the Turing Award, now considered the equivalent of the Nobel Prize in the computing world.

Vallée-Poussin, Charles Jean de la (1866–1962): This Belgian mathematician independently produced a proof of the prime number theorem in 1896, the same year that Jacques Hadamard also produced a proof.

Wiles, Andrew (1953–): This British mathematician on the Princeton faculty worked for nearly 10 years to prove a result involving elliptic curves that finally proved Fermat's Last Theorem in 1993. He has received many awards, including the Cole Prize and Wolf Prize.

Bibliography

Burger, Edward B. *Student Mathematical Library. Vol. 8, Exploring the Number Jungle: A Journey into Diophantine Analysis.* Providence: American Mathematical Society, 2000. Focuses on discovering the hidden truths and treasures of Diophantine analysis.

———. *Zero to Infinity: A History of Numbers.* Chantilly, VA: The Teaching Company, 2007. A 24-lecture video of the history of numbers that sets up the background for this course.

Burger, Edward B. and Robert Tubbs. *Making Transcendence Transparent: An Intuitive Approach to Classical Transcendental Number Theory.* Cambridge: Springer, 2004. An introduction to the challenging field of transcendental number theory.

Hardy, G. H. and E. M. Wright. *An Introduction to the Theory of Numbers.* 5th ed. New York: Oxford University Press, 1980. A good overall introduction.

Niven, Ivan, Herbert S. Zuckerman, and Hugh L. Montgomery. *An Introduction to the Theory of Numbers.* New York: John Wiley & Sons, Inc., 1991. Another good overall introduction.

Stark, Harold M. *An Introduction to Number Theory.* Cambridge: MIT Press, 1978. A useful overview.

Answers to Selected Questions to Consider

Lecture One

1. Courtroom proceedings, crafting legislation, contract negotiations.

2. Argue by contradiction: Suppose the number 21 is *not* interesting. Then, because we are supposing that the numbers 1 through 20 *are* interesting, we find that 21 is the smallest *un*interesting number. But this makes 21 interesting! We have a contradiction to our assumption about 21, and thus 21 must be interesting.

Lecture Two

1. The number 64 is a perfect square: $64 = 8 \times 8$. It is also a perfect cube: $64 = 4 \times 4 \times 4$. In fact, 64 is a power of 2: $64 = 2 \times 2 \times 2 \times 2 \times 2 \times 2 = 2^6$. The number 61 is neither a perfect square nor cube. It does not have 2 as a factor and so is odd. Not only that, 61 has no factors other than itself and 1.

2. Following the algorithm in the lecture, we generate the following sequence: 7, 22, 11, 34, 17, 52, 26, 13, 40, 20, 10, 5, 16, 8, 4, 2, 1, and now it becomes periodic.

Lecture Three

1. Using our formula, we have:
$(1{,}000{,}000 \times 1{,}000{,}001)/2 = 500{,}000{,}500{,}000$.

2. Look the table below for values for several pairs of consecutive triangular numbers. Notice the amazing pattern!

Triangular numbers	Squares	Differences of squares	Perfect cubes!
1	1		
3	9	$9 - 1 = 8$	$8 = 2^3$
6	36	$36 - 9 = 27$	$27 = 3^3$
10	100	$100 - 36 = 64$	$64 = 4^3$
15	225	$225 - 100 = 125$	$125 = 5^3$
21	441	$441 - 225 = 216$	$216 = 6^3$

Lecture Four

1. The suggested hint points out that this progression can be obtained from the progression 1, 5, 25, 125, … after we multiply each term by −2. So we will first find the sum of the first five terms of the progression 1, 5, 25, 125, … . Using the notation from the lecture, we notice that this progression has $r = 5$, so the sum of the first five terms is $(5^5 − 1)/(5 − 1) = 3124/4 = 781$. We now multiply this answer by −2 to find that the sum of the first five terms of our original progression is −1562.

2. We notice that the given infinite geometric series can be obtained from the infinite geometric series $1 + 1/10 + 1/100 + 1/1000 + \cdots$ after we multiply the entire sum by 9/10. So we will first find the sum of the infinite series $1 + 1/10 + 1/100 + \cdots$, which has an r value equal to 1/10. The formula in the lecture gives us a sum of $1/(1 − r) = 1/(1 − 1/10) = 10/9$. We now multiply this answer by 9/10 to find that the sum of our original progression is 1. Thus we have shown that 0.999… = 1. Amazing (and correct)!

Lecture Five

1. a) The pattern explored in the lecture suggests that the sum of the first n Fibonacci numbers is 1 less than the Fibonacci number two steps further down the list. So the sum of the first 10 Fibonacci numbers should be 1 less than the 12th Fibonacci number, which is 144. Thus our desired sum equals 143.

    b) The largest Fibonacci number less than 100 is 89, so we have 100 = 89 + 11. We see easily that 11 = 8 + 3, two more Fibonacci numbers, giving us 100 = 89 + 8 + 3. We replace each term in this sum with its Fibonacci successor (a number approximately 1.6 times as large) to get 144 + 13 + 5 = 162. So 100 miles is approximately 162 kilometers.

2. N/A

Lecture Six

1. This sequence begins with seed values 1 and 1. Successive terms are obtained as follows: Multiply the most recent term by 3 and add the result to the term that came before. Thus $4 = 3 × 1 + 1$, $13 = 3 × 4 + 1$, $43 = 3 × 13 + 4$, and so on.

2. We use the general formula given in the lecture: The puzzle with n disks requires $2^n - 1$ moves. With $n = 10$, we need $2^{10} - 1 = 1023$ moves.

Lecture Seven

1. We consider the product $2 \times 3 \times 5 \times 7 \times 11 \times 13 \times 17 \times 19$ and then add 1. Because this number has remainder 1 when divided by each prime, 2, 3, 5, 7, 11, 13, 17, and 19, and because 20 itself is not a prime, any prime factors of this large product (plus 1) must be greater than 20. But we know this number is either itself prime or can be written as a product of primes, hence in either case there must exist a prime greater than 20.

2. The number $101 \times 100 \times 99 \times 98 \times 97 \times \cdots \times 3 \times 2 \times 1$ can be written compactly as "101!" (read "101 factorial"). Clearly the numbers 2, 3, 4, ... , 101 are factors of 101!. Thus we observe that 101! + 2 has 2 as a factor, 101! + 3 has 3 as a factor, 101! + 4 has 4 as a factor, and so on, up to 101! + 101, which has 101 as a factor. Therefore we have 100 consecutive natural numbers: 101! + 2, 101! + 3, 101! + 4, $\cdots$, 101! + 101, each of which is composite. (Note: We do not claim that this list is the *smallest* 100 consecutive composite numbers, but it is a list of 100 consecutive numbers that we are *certain* are all composite!)

Lecture Eight

1. Because every 7th natural number is a multiple of 7, the probability that a natural number chosen at random is a multiple of 7 is 1/7. Thus, the probability that a natural number is *not* a multiple of 7 is $1 - 1/7 = 6/7$.

2. We must sum the first 4 terms in the series to get a partial sum greater than 2, and we must sum 11 terms to exceed 3.

Lecture Nine

1. No such progression of primes exists. Here is why: If we start with $n = 2$, then $n + 3 = 5$, which is prime, but $n + 6 = 8$, which is not. All other primes are odd. But if n is odd, then $n + 3$ would be an even number greater than 2 and thus cannot be prime.

2. We produce Fermat primes for $n = 1$ and $n = 2$ as follows: $2^{2^1} + 1 = 2^2 + 1 = 5$ and $2^{2^2} + 1 = 2^4 + 1 = 17$. (Note that when $n = 3$, the formula yields 129, which has a factor of 3 and therefore is not prime.)

Lecture Ten

1. Because the number 123,456,789 − 213 has a factor of 123, we know that the numbers 123,456,789 and 213 must have the *same* remainder when divided by 123. So to answer the question, we take the easy way out and simply divide 123 into 213. We get a quotient of 1 and a remainder of 90, which must also be the remainder after dividing 123,456,789 by 123.

2. N/A

Lecture Eleven

1. N/A

2. Because 29 has no common factors with 31, Fermat's Little Theorem tells us that $29^{30} \equiv 1 \pmod{31}$, so the remainder is 1 (notice that the exponent 30 equals 31 − 1).

Lecture Twelve

1. We note that of all the possible remainders when dividing by 12: 0, 1, 2, 3, 4, 5, 6, 7, 8, 9, 10, and 11, only 4 of them are relatively prime to 12: 1, 5, 7, and 11. So by the extension of Fermat's Little Theorem due to Euler, because 7 is relatively prime to 12, we have $7^4 \equiv 1 \pmod{12}$.

2. N/A

Lecture Thirteen

1. Suppose the integers x, y, and z formed a solution to the equation. The left side would be an integer that when divided by 3 gives a remainder of 0. But the right side would be an integer that gives a remainder of 2 when divided by 3. This is impossible, so no such integer solution exists.

2. Fermat's method of ascent as applied to this equation gives formulas for obtaining a new solution from an old one: $x_{new} = 3x_{old} + 4y_{old}$ and $y_{new} = 2x_{old} + 3y_{old}$. These formulas were used in the lecture to generate the solution $x = 17$ and $y = 12$ from the solution $x = 3$ and $y = 2$. We now substitute $x_{old} = 17$ and $y_{old} = 12$ into the formulas to obtain a third solution: $x_{new} = (3 \times 17) + (4 \times 12) = 99$ and $y_{new} = (2 \times 17) + (3 \times 12) = 70$. It is easy to check that $99^2 - (2 \times 70^2) = 1$, so this new solution does work.

Lecture Fourteen

1. Suppose, contrary to what we wish to establish, that we had a triple of natural numbers x, y, and z that satisfied $x^{100} + y^{100} = z^{100}$. This equality is equivalent to the equation $(x^{20})^5 + (y^{20})^5 = (z^{20})^5$, giving us natural numbers x^{20}, y^{20}, and z^{20} satisfying the equation $x^5 + y^5 = z^5$. This is a contradiction since we know that it has been shown that there are no natural numbers that satisfy this Diophantine equation, so our assumption is false, and therefore no natural-number solution to $x^{100} + y^{100} = z^{100}$ exists.

2. The next Germain prime is 23. We note that $2 \times 23 + 1 = 47$, which is indeed prime. It is easy to check that none of the primes between 11 and 23 are Germain primes. For example, 13 is not a Germain prime because $2 \times 13 + 1 = 27$, which is not prime.

Lecture Fifteen

1. There are two ways to factor 100 into "primes" in the ring of even integers: $100 = 10 \times 10$, and $100 = 2 \times 50$.

2. N/A.

Lecture Sixteen

1. The length you measure should equal 5 inches. You are measuring the sides of what should be a right triangle, so the length of the hypotenuse squared should be the sum of the squares of the other two lengths, in this case, $3^2 + 4^2 = 25$.

2. Because this method generates a different Pythagorean triple for each natural number greater than 1 and there are infinitely many such numbers, you have shown that there are infinitely many such triples.

Lecture Seventeen

1. The point $(-1, 0)$ is on the line because substituting $x = -1$ and $y = 0$ into equation $y = 1/2(x + 1)$ gives a valid equation. To find the second point where this line intersects the unit circle, we use the formulas derived in the lecture: $X = (1 - m^2)/(1 + m^2)$, and $Y = 2m/(1 + m^2)$. The value of m is the slope of our line; in this case $m = 1/2$. Substituting into our formulas, we obtain $X = (1 - (1/2)^2)/(1 + (1/2)^2) = (1 - 1/4)/(1 + 1/4) = 3/5$ and

$Y = 2(1/2)/(1 + (1/2)^2) = 1/(1 + 1/4) = 4/5$. So the second point is $(3/5, 4/5)$.

2. Substituting $x = 8/17$ and $y = 15/17$ into $x^2 + y^2 = 1$, we get $(8/17)^2 + (15/17)^2 = 64/289 + 225/289 = 289/289 = 1$. Thus the given point does lie on the unit circle. The corresponding Pythagorean triple is $(8, 15, 17)$.

Lecture Eighteen

1. For the point $(1, 1)$, we substitute $x = 1$ and $y = 1$ into the equation to obtain $1 = 1 - 1 + 1$, which is valid. For the point $(0, 1)$, we substitute $x = 0$ and $y = 1$ to obtain $1 = 1$, which is also valid.

2. Setting $y = 0$, the equation becomes $0 = x^3 - 4x$. The right side factors to yield $0 = x(x^2 - 4) = x(x - 2)(x + 2)$. The three factors on the right multiply to give 0, so one of them must be 0. Thus we have $x = 0$, $x = 2$, or $x = -2$.

Lecture Nineteen

1. We suppose $\sqrt{3}$ is rational and work toward a contradiction. If $\sqrt{3}$ is rational, then $\sqrt{3} = a/b$ for some integers a and b. So $3 = a^2/b^2$, and thus $3b^2 = a^2$. Recall that every natural number can be written uniquely as a product of primes. Note also that 3 is prime and that 3 must appear an even number of times in the prime factorizations of a^2 and b^2. But then the equation $3b^2 = a^2$ would have an odd number of 3s dividing the left side and an even number of 3s dividing the right side, which is impossible. Thus our original assumption must have been faulty, and so $\sqrt{3}$ is irrational.

2. Given that α is an irrational number, we know that its decimal expansion never terminates or becomes periodic. The decimal expansion for 10α is obtained from the expansion for α by moving the decimal point one digit to the right. Thus 10α also has a non-terminating, non-periodic expansion, and therefore it must also be irrational.

Lecture Twenty

1. We have $4x = 0 - 5$, so $4x = -5$, and thus $x = -5/4$.

2. We know that raising $1/\sqrt[3]{2}$ to the third power yields $1/2$. Therefore $x = 1/\sqrt[3]{2}$ satisfies the equation $x^3 - 1/2 = 0$. To obtain integer coefficients, we multiply through by 2 to obtain $2x^3 - 1 = 0$.

Lecture Twenty-One

1. The first such power is 46: $2^{46} = 70{,}368{,}744{,}177{,}664$. (Please note that this may be difficult to verify without a calculator or software that carries at least 14 significant digits.)

2. Comparing the decimal expansions of 22/7 and 31/10 to that of π:

    $$\pi = 3.14159265\ldots$$
    $$22/7 = 3.14285714\ldots$$
    $$31/10 = 3.10000000\ldots$$

    we see that 22/7 is accurate to two decimal places, whereas 31/10 is accurate to only one. Dirichlet's Theorem from the lecture also tells us that 22/7 lies within 1/56 of π. In addition, 22/7 uses a smaller denominator than 31/10, and so is "less expensive."

Lecture Twenty-Two

1. We compute:

    $$\frac{18}{7} = 2 + \frac{4}{7} = 2 + \frac{1}{7\big/4} = 2 + \frac{1}{1 + 3\big/4} = 2 + \cfrac{1}{1 + \cfrac{1}{4\big/3}} = 2 + \cfrac{1}{1 + \cfrac{1}{1 + \cfrac{1}{3}}}.$$

2. Using a calculator, we find $e = 2.7182818284\ldots$. Thus:

    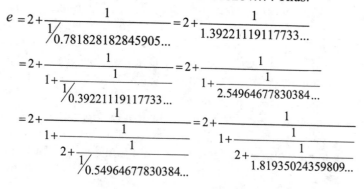

    and continuing in this manner reveals the first few partial quotients.

1. We note that: $\sqrt{5} = 2 + \cfrac{1}{4 + \cfrac{1}{4 + \cfrac{1}{4 + \cfrac{1}{\ddots}}}}$.

The next three best rational approximants are:

$$2 + \frac{1}{4} = \frac{9}{4}, \ 2 + \cfrac{1}{4 + \cfrac{1}{4}} = 2 + \cfrac{1}{17\!\!\diagup\!\!4} = 2 + \frac{4}{17} = \frac{38}{17}, \text{ and}$$

$$2 + \cfrac{1}{4 + \cfrac{1}{4 + \cfrac{1}{4}}} = 2 + \cfrac{1}{4 + \cfrac{1}{17\!\!\diagup\!\!4}} = 2 + \cfrac{1}{4 + \cfrac{4}{17}} = 2 + \cfrac{1}{72\!\!\diagup\!\!17} = 2 + \frac{17}{72} = \frac{161}{72}$$

2. As described in the lecture, we look at each rational approximant as a fraction x/y. So first we have $x = 9$ and $y = 4$. Substituting into the given Pell equation, we see $(9)^2 - 5(4^2) = 81 - 80 = 1$, so $x = 9$ and $y = 4$ is a solution. Because we learned that every other approximant is a solution n, we now check $x = 161$ and $y = 72$ to find $(161)^2 - 5(72^2) = 25{,}921 - 25{,}920 = 1$. Thus $x = 161$ and $y = 2$ is also a solution.

1. N/A.

2. N/A.

Notes